INSTRUMENTATION
& CONTROL SYSTEMS
ENGINEERING HANDBOOK

No. 1035
$19.95

INSTRUMENTATION & CONTROL SYSTEMS ENGINEERING HANDBOOK

BY THE EDITORS OF
INSTRUMENTATION TECHNOLOGY

TAB BOOKS

BLUE RIDGE SUMMIT, PA. 17214

FIRST EDITION

FIRST PRINTING—DECEMBER 1978

Library of Congress Cataloging in Publication Data

Main entry under title:

Instrumentation & control systems engineering handbook.

    Includes index.
    1. Automatic control. 2. Measuring instruments. 3. Systems engineering. I. Instrumentation technology.
TJ213.I54   628.8'3   78-11391
ISBN 0-8306-9867-1

Contents

Section IV

Foreword

Anyone specializing in instrumentation and controls realizes the rapid expansion of knowledge in these areas. Skills learned soon become obsolete. And new methods continually displace old ones—as with the rapid extension of computers into instrumentation and control functions. Attempting to keep up with the mass of literature devoted to these diverse skills is a major chore for professionals. The Instrument Society of America, through its journal *Instrumentation Technology* and other media, keeps its members up to date with the latest developments.

This book collects information of lasting interest from *Instrumentation Technology*. This information falls into four broad categories: process analysis, measurement, control valves, and control electronics. Theoretical guidance as well as specific details of methods and equipment is given. Theoretical information paints the broad strokes intended to evoke a fresh look at old problems. For instance, "Fiber Optics for Data Transmission" gives a detailed comparison of this newcomer to older methods of data transmission. Solutions to specific problems also are suggested. For example, "Measuring Torsional Vibration" deals with problems of rotating mechanisms.

Even while rapidly expanding, the field must not only meet the old demands of science and technology but also requirements like those of the Environmental Protection Agency (EPA) and the Occupational Safety and Health Administration (OSHA). The EPA, for example, requires continuous stack monitors on many power boilers, and OSHA enforces noise standards for workplace equipment. These diverse requirements place multiple demands on those involved in the complex world of instrumentation and controls.

In recognition of this greatly expanded role, the Instrumentation Society of America recently revised their constitution to read: "The objectives of the society shall be to advance and to reinforce the arts and sciences related to the theory, design, manufacture and use of *instrumentation, computers and systems for measurement and control* in the various sciences and technologies for the benefit of mankind." The italicized words replace the words *instruments and controls*. This revision did not transpire in anticipation of future developments; it merely reflected conditions as they had developed.

Professionals must attempt to keep up with this mushrooming body of knowledge. But if anyone set out to attend all the manufacturer's training programs, university short courses, meetings and symposiums, and various shows, little (if any) time would be left for applying the learning. And only by applying what has been learned can people expand their proficiency. Knowledge not put into practice soon dissipates, like vapor from an open container.

The value of a work such as this lies in its availability. Resting on a bookshelf, it waits for the times when needed. For research during initial design stages, there is information in "Selecting the Right Flow Meter," "Review of Temperature Measurement Techniques," and "Humidity Measurement." Deep into a project, a designer might be interested in information devoted to particular problems. Look in "Commonmode Rejection Techniques for Low-level Data Acquisition," "Automatic Bridge Balancing Circuit," "Accuracy in Signal Conversion," and "Solid-state vs Electromechanical Relays."

But to wait until some problem crops up before looking for a solution isn't the most efficient use of this, or any, book. Ideally, a user should know what is available between the covers and refer to it as necessary. Information follows a logical pattern. Introductory paragraphs establish the theme or present the problem. Subsequent paragraphs then methodically develop the presentation through to the end.

Reference lists are included for the most part. This allows those interested in obtaining more information to pursue the subject further.

This book can be used as a self-contained course in instrumentation and controls or for answers to specific problems. In either case, the book should prove to be a handy reference on your shelf.

Editors, *Instrumentation Technology*

Section I—Process Analysis

This first section contains information pertaining to process analysis. Leading off the section is a two-part treatment of troubleshooting process control loops. Part I discusses maintaining stability and accuracy. Part II delves into transients and dynamic compensation.

The next two parts consider the dynamic design of pneumatic control loops. They develop analysis by segments and then application principles and practices. Sixteen references are listed for those interested in learning more about the author's presentation.

Wrapping up the section is information about batch control. An introduction to the dynamics and control of batch chemical reactors is presented first. The next two parts discuss batch-control problems. These focus first on improving process controllability with unconventional measurement and control methods. They then consider hardware and software parameters that influence system safety and reliability. The author concludes by outlining future trends in batch control systems.

Troubleshooting
Process Control Loops

C. H. CHO

PART I: MAINTAINING STABILITY AND ACCURACY

Unexpected control problems in process applications generally become evident during plant startup or when plant capacity is changed to a new value. At these times, the process engineer may find that the performance characteristics of certain loops no longer meet the original control specifications. However, it is not always convenient for him to test and evaluate problem loops throughout their operating range and to make certain that their controllers are properly tuned to handle any unforeseen gain change, particularly at startup.

Applying control theory to such situations can help to identify the loop components causing the out-of-specification condition so that they can be modified or replaced. A number of closed loop control problems can be avoided if the system designer takes the time to synthesize the system and select components which will satisfy the requirements of the specific process. He can perform a simple open-loop analysis using Bode or Nyquist stability criteria to determine the static and dynamic requirements of each hardware element in the alternative system or base his component selection on well-established guidelines.

To gain a good understanding of expected loop performance, it is important for the control engineer to know both the static and dynamic characteristics of each control component in the loop. This information, along with process dynamics data, gives the control engineer a tool with which he can make a quick system analysis in the frequency domain.

Every control loop has a particular function in terms of the control objectives which the system designer wishes to obtain. Most control loops are subjected to at least two types of inputs: 1) a command signal for setpoint control and 2) a load input which creates a transient or a steady-state setpoint deviation. When any one of the following system performance criteria is found to be unacceptable, the process engineer must troubleshoot the loop to find the problem component and modify its characteristics so that the loop will meet control specifications.

Stability—The most important performance criterion of any control loop is its stability through the operating range of the system. The loop must be stable in response to a load disturbance.

Setpoint control accuracy—Loop specifications generally state the acceptable deviation from the setpoint in terms of some percentage of the full-scale value of the controlled variable.

Initial overshoot—In some applications, a large overshoot cannot be tolerated because of the physical stress limits of pressure vessels or pipes. In blending and batch applications, control of overshoot is important because of its potential effect on the composition of the final product and is allowable variation as a part of the loop specification.

Transient response—The transient response characteristic of a loop as a function of load or setpoint change is an important control parameter. The time to the first overshoot (T_p) and settling time (T_s) in response to a unit step change are a part of the overall response characteristics.

Performance criteria for transient response are particularly important in applications where the test duration may be limited to 20 s or less. For example, jet engine tests and wind tunnel operations both involve a tight specification on the time required to establish and maintain the control point close to its designed setpoint throughout the test. In process control applications, operating personnel are quite concerned with wear of the final control elements if the system must undergo a long, sustained oscillation before it reaches its steady-state value. In addition, such variations may have an undesirable effect on the quality of the product.

The basic objective of any troubleshooting is to locate and clearly define the loop problem. A prerequisite is to answer the following questions: 1) What type of process is being controlled? 2) What does the system schematic look like? 3) What is the controlled variable? 4) What are the control mode settings? 5) Is there any recorded information available, or can the results of the loop problem be seen? 6) Is the system stable on open loop? 7) If the system is cycling, what is the cycling frequency?

All of the possible process control problems which can necessitate troubleshooting cannot be covered adequately here. Since control engineers are typically concerned with performance-related specs when synthesizing or analyzing control systems, the remainder of this article concentrates on the performance factors listed above. Stability, steady-state control accuracy and transient response are the primary factors to be considered in troubleshooting feedback loops. Figure 1-1A shows a single-loop process control system which forms the basis for discussion.

LOOP STABILITY PROBLEMS

Knowing the dynamic characteristics of the selected control components and the process, it is possible to evaluate the stability of a system theoretically. However, the most essential part of systems troubleshooting is a good understanding of the process in terms of its static and dynamic characteristics, because this knowledge dictates the selection of alternative control equipment.

The Bode stability criterion states that the open-loop gain of a feedback system, such as the one shown in Fig. 1-1A, must be less than one at its critical frequency. This can be stated mathematically as follows:

$$(K_c D_c)\ (K_a D_a)\ \ (K_p D_p)\ (K_t D_t) < 1 \qquad (1.1)$$

where K represents a static gain and D a dynamic gain; the subscripts c, a, p and t denote the controller, actuator/valve, process and transmitter, respectively.

Figure 1-1B shows the behavior of this closed-loop system for the three possible value conditions of open-loop gain in response to a setpoint or load disturbance. Depending on the open-loop gain at the critical frequency (the frequency at which a 360-degree phase shift occurs in the open loop), the system will exhibit an oscillation, or converging oscillation. Only the last response is a stable one.

If a system is unstable, its controlled variable will either diverge from the setpoint or maintain a continuous oscillation about it. In the latter case, where open-loop gain equals 1, the system response maintains a fixed amplitude because of such constraints as physical limits to the valve travel distance. The steps which follow are recommended to identify the source of instability in a feedback configuration.

Put the controller on manual to see if the system is stable in the open-loop condition. If the output signal still oscillates, the loop may contain a control component which becomes unstable in the presence of certain load signals. Control components which have such nonlinear characteristics as deadband, saturation or backlash can cause the output to oscillate at a constant amplitude; this behavior is described as a limit

16

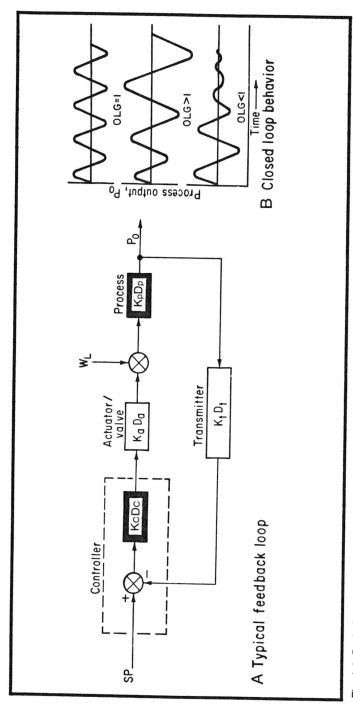

Fig. 1-1. Basic feedback loop (A) exhibits varying degrees of stability in its response (B), depending on the value of the open-loop gain (OLG) at the critical frequency. OLG is the product of gains of the individual components, each of which has a static (K) and a dynamic (D) factor.

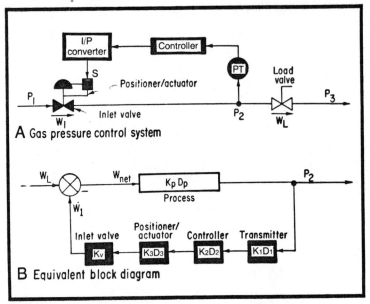

Fig. 1-2. The process dynamics of a gas pressure control system (A) are represented by the equivalent block diagram (B). The system is designed to maintain gas header pressure P_2 (output) constant, in spite of variations in supply pressure P_1 or gas flow demand W_L.

cycle. Such devices are generally sensitive to input amplitude and/or frequency.

The only way the troubleshooter can isolate the source of the limit cycle problem is to check the input and output of each individual component with test equipment, having a means of manually overriding the normal input signal with the test input signal. The limit cycle frequency in most process control applications is considerably faster than the loop cycling frequency.

The most common cause of loop instability—the one indicated when the manual test shows that the open loop is stable—is a change in the static and dynamic characteristics of the process as a function of the load. This effect can be described in terms of a typical gas pressure control system, Fig. 1-2A, whose equivalent block diagram is shown in Fig. 1-2B. The process dynamics of this system can be derived with the following assumptions: inlet pressue P_1 and outlet pressure P_3 are constant; the piping system can be represented as a lumped parameter system; the pressure drops across the inlet control valve and the load valve are both subcritical.

A self-regulation term for the entire process can be derived from any suitable compressible flow equation by taking the partial derivative of the flow with respect to the controlled pressure P_2. The value of this term expresses the small perturbation of gain around a given operating point.

The flow equation selected is Fliegner's empirical flow equation for compressible fluids (R. G. Hudson, *Engineers Manual*, John Wiley & Sons, New York, p. 166):

$$W = \frac{1.06 \, A \, K_f \sqrt{P_2(P_1 - P_2)}}{\sqrt{\theta_1}} \qquad (1.2)$$

where

W = mass rate of flow, lbs/s
A = area of valve opening, in.^{2}
K_f = valve flow coefficient
P_1 = upstream pressure, psia
P_2 = controlled pressure, psia
P_3 = downstream pressure, psia
θ_1 = upstream temperature, °R

Transient flow responses are considered in two parts: across the inlet valve (W_1) and across the load valve (W_L). Taking the partial derivative of W_1 with respect to P_2 and combining the result with Equation 1.2 reduces to:

$$\frac{\delta W_1}{\delta P_2} = \frac{W}{P_2} \left[\frac{1 - 2P_r}{2(1 - P_r)} \right] \qquad (1.3)$$

where $P_r = P_2/P_1$ and $P_r > 0.5$

Similarly, taking the partial derivative of flow across the load valve W_L with respect to P_2 yields:

$$\frac{\delta W_L}{\delta P_2} = \frac{W}{P_2} \left[\frac{1}{2(1-P_{r1})} \right] \qquad (1.4)$$

where $P_{r1} = P_3/P_2$.

The total process self-regulation ψ is the sum of Equations 1.3 and 1.4:

$$\psi = \frac{W}{P_2} \left[\frac{1}{2(1-P_{r1})} + \frac{1-2P_r}{2(1-P_r)} \right] \qquad (1.5)$$

The transfer function of the process can then be approximated as a first-order lag:

$$\mathcal{L}\ (P_2/W_L) = \frac{1/\psi}{\left(\dfrac{C}{\psi}\right) s + 1} \qquad (1.6)$$

where

C = pneumatic capacitance = V_g/a^2
and V = control volume, in.^{3}

g = gravitational constant, in./s^2

a = sonic velocity, in./s

Figure 1-3 illustrates the frequency response plot corresponding to Equation 1.6. The magnitude curves (top) show that process gain K_p varies in a nonlinear manner as a function of the load W_L on the system. The generalized plot of the static process gain, Fig. 1-4, indicates that K_p should be compensated to obtain satisfactory system stability under all load conditions.

Rewriting Equation 1-1 by substituting the process transfer function (Equation 1.6) for the gain elements K_pD_p, the open-loop gain (OLG) of the system shown in Fig. 1-2 can be expressed as:

$$OLG = \frac{1/\psi}{\left(\dfrac{C}{\psi}\right) s + 1} (K_1D_1)(K_2D_2)(K_3D_3)(K_v) \qquad (1.7)$$

where the valve dynamics are assumed negligible ($D_v = 1$). Since ψ is a function of W (Equation 1.5), Equation 1.7 shows that the open-loop gain is a function of the load. Assuming all other loop components are linear, this nonlinearity in process gain K_p is plotted in Fig. 1-4. Note that $(K_p)\alpha > (K_p)\beta > (K_p)\gamma$. The loop must either be compensated for this nonlinearity or it should be tuned for a process gain of $(K_p)\alpha$ (under the low flow rate condition). Thus, when the operating point changes to 50 percent or 100 percent of the full load, the system can still operate in a stable manner, although with a somewhat poorer transient response. If the ratio of $(K_p)\alpha$ to $(K_p)\gamma$ is greater than 2, and the loop is tuned for a process static gain of $(K_p)\gamma$ with a margin of 2, the system will become unstable when the load demand changes from 100 to 25 percent of the full load condition.

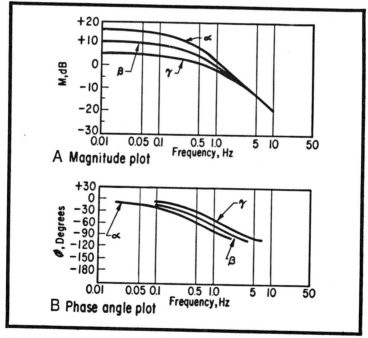

A Magnitude plot

B Phase angle plot

Fig. 1-3. Frequency response plots of the gas pressure control system of Fig. 1-2 show nonlinear process gain variations under low (α), medium (β) and high (γ) flow rate conditions. As load flow rate W_L increases from α to β to γ, process gain K_P decreases.

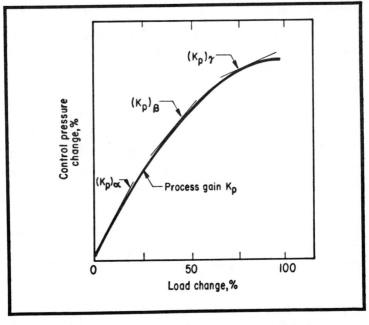

Fig. 1-4. This plot of open-loop static gain K_p of the gas pressure control vs load shows the nonlinearity. Process gain at a given load value is equal to the slope of the curve at that point. Stability can be preserved compensating the nonlinearity with an equal percentage valve or by tuning the loop so that, under low flow conditions, the loop gain is equal to 1 at the system's critical frequency with a process static gain of $(K_p)\ \alpha$.

HARDWARE COMPENSATION OPTIONS

In terms of the set of system operating conditions selected for this gas pressure control example, an equal percentage valve can compensate for the process nonlinearity.

In fact, the control valve should be viewed as a most effective means of nonlinear compensation for process control applications which have a nonlinear element in the loop. Most control valve manufacturers do make a serious attempt to provide at least three basic inherent control valve characteristics. These are linear, equal percentage and quick opening characteristics. The basic equation governing the control valve gain can be written:

$$K_v = n\ \left[\frac{W_{max}}{Y_{max}}\right] \qquad (1.8)$$

22

where

W_{max} = maximum flow based on maximum C_v (liquid valve size coefficient)

Y_{max} = maximum valve lift

n = slope of the percent C_v -vs- percent Y plot at a given operating point

Each valve characteristic has an associated n value: for quick-opening valves (over the lower 70 percent of the flow range), $n = 1.6$; for linear valves, $n = 1$; for equal percentage valves,

$$n = \frac{\% \, C_v(lnR)}{100} = \frac{W \, (lnR)}{W_{max}} \qquad (1.9)$$

where R = inherent rangeability. Combining Equations 1.8 and 1.9:

$$K_v = \frac{W \, (lnR)}{Y_{max}} \qquad (1.10)$$

From Equation 1.5, the static gain of the process, $1/\psi$, can be expressed as NP_2/W, where N is the reciprocal of the

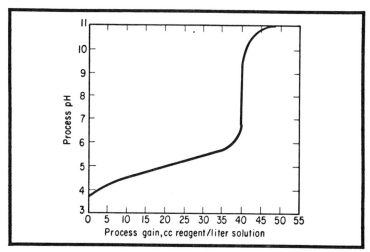

Fig. 1-5. A difficult control problem may arise in a pH feedback loop if the control point is set at the point of neutrality (7 on the pH scale) with a standard PID controller. Process gain on both sides of neutrality is very low, whereas in the vicinity of pH 7, it is extremely high. A wide range of gain perturbations along this curve can be compensated by a nonlinear controller having an adjustable notch gain.

bracketed terms. Substituting this and Equation 1.10 into Equation 1.7 yields:

$$OLG = NP_2(K_1D_1) \ (K_2D_2)(K_3D_3)(lnR)/Y_{max} \quad (1.11)$$

Thus, the flow variable W has been cancelled by inserting an equal percentage valve characteristic (Equation 1.10) into the loop. The open-loop static gain then remains constant throughout its operating range.

Placing a cam in the feedback path of the valve positioner/actuator is another effective means of generating the type of function needed to compensate for nonlinearities. However, the bandwidth of the positioner/actuator should be a factor of 10 greater than the critical frequency of the system for effective compensation. This gas pressure control system could also be compensated by inserting an equal percentage cam in the positioner/actuator of a valve having a linear characteristic.

In pH control applications, the process typically exhibits a very high gain around the point of neutrality, as indicated in Fig. 1-5. Titration curves for many different types of pH processes have been well-defined by process engineers, and such characteristic curves show whether pH control at a given setpoint will be straightforward or difficult. As shown in Fig. 1-5, the process gain of 40 cc of reagent per liter solution is extremely high, whereas the gain on both sides of this region of neutrality is very low. Therefore, pH control hardware must be designed with features and gain adjustments that meet the stability requirements for a given pH setpoint along this curve. A nonlinear controller with an adjustable notch gain and normal proportional gain outside of the notch makes a good pH controller, but this is a difficult task for a standard linear PID controller.

If the loop stability problem can be attributed to the nonlinear, discontinuous behavior of a particular component, an understanding of the nature of the nonlinearity and its describing function is necessary so that an analysis can be made to determine the sensitivity relationships among: 1) the

24

input amplitude and/or frequency, 2) the magnitude of the nonlinearity, 3) the linear gain and 4) the system's limit cycle frequency and amplitude. If the limit cycle frequency is sufficiently higher than the system's critical frequency and its amplitude is small enough, many process control systems can tolerate its presence. However, there are many practical ways to achieve stability such as simply biasing the backlash in gear trains with a spring or reducing valve deadband with better packing and sealing material. However, these approaches may not always be practical or even possible.

Misapplication of controller modes can cause difficult control problems. For example, in fast and noisy processes such as liquid pressure, small capacitance gas pressure, and most liquid level control systems, the derivative mode can amplify process noise; the loop generally experiences random oscillation and instability problems as well. Furthermore, the derivative mode is completely ineffective in introducing phase lead and high frequency gain boost for fast process loops because of the ultimate frequency response limitation of most process controllers. The bandwidth of most controllers is 15 Hz or less.

STEADY-STATE ACCURACY PROBLEMS

Steady-state control accuracy, the error between the setpoint and the controlled variable, is inversely proportional to the open-loop static gain of a system. Referring again to Fig. 1-1, the relationship between the load change and the output change can be expressed by the closed-loop transfer function:

$$\mathcal{L}\,(P_o/W_L) = \frac{K_pD_p}{1 + K_pD_pK_cD_cK_aD_aK_tD_t} \quad (1.12)$$

Steady-state accuracy can be evaluated by modifying Equation 1.12, assuming that the static feedback gain is much greater than one ($K_cK_aK_t >> 1$):

$$\mathcal{L}\,(P_o/W_L) = \frac{1}{K_cK_aK_t} \quad (1.13)$$

$$s \to 0$$

25

If the term $K_cK_aK_t$ approaches infinity, the controlled variable P_o becomes completely insensitive to load changes insofar as the steady-state error is concerned. However, an obvious conflict exists between obtaining the maximum loop gain which the system can tolerate to achieve high accuracy, without at the same time causing instability. Incorporating the integral mode into the controller usually alleviates this conflict. However, remember that the integral mode adds a phase lag of—90 degrees, which significantly influences the critical frequency and the stability of the system. Obviously, the maximum limit on gain is dictated by the critical frequency and gain margin at that frequency.

Most PI controllers have a reset gain of about 200, which permits use of the integral mode for good control accuracy. The low frequency integral gain of the controller should be analyzed with either frequency response or step test inputs to compare the actual response against the manufacturer's specification. Not only is the reset gain important, but also the range of reset time with which the control engineer can establish the reset corner. This frequency corner should be lower than the system's critical frequency, by a factor of at least two or three.

The accuracy with which the instrument engineer can establish a setpoint depends on the static accuracy of the process measuring device. Therefore, it is important to select a transmitter which has a good static accuracy. The closed-loop transfer function between a setpoint change and the process variable is:

$$\mathcal{L} (P_o/SP) = \frac{K_cD_cK_aD_aK_pD_p}{1 + K_cD_cK_aD_aK_pD_pK_tD_t} \quad (1.14)$$

and the steady-state accuracy can be evaluated from:

$$\mathcal{L} (P_o/SP) = \frac{1}{K_t} \quad (1.15)$$
$$s \to 0$$

The feasibility of establishing precise setpoint control

also depends on the setpoint calibration of the controller and its accuracy. The same consideration applies to setpoint accuracy if the controller accepts a remote setpoint signal from another controller (cascade control system). Therefore, there are two sources of errors which may affect setpoint control accuracy: the sensor/transmitter and the controller itself. If the problem of setpoint accuracy arises, then both devices must be calibrated and checked against hardware performance specs.

PART II: TRANSIENTS AND DYNAMIC COMPENSATION

Disturbances, load changes and other upsets are common occurrences in any process plant operation; and the transient response of the controlled plant to such varying conditions, as well as to setpoint changes, is an important part of the control system specifications. Mathematically, the transient response can be determined by taking the inverse transform of the closed-loop process transfer function, but this procedure is a very tedious one, even for low-order systems.

Instead of a mathematical transformation from the Laplace or s-domain to the time domain, transient response can be assessed much more simply by a graphical technique in the s-plane. A figure of merit for system response can be calculated with a simplified method that utilizes a pole-zero diagram. This method graphically relates characteristics of the transient response to the pattern of poles and zeros of the control ratio for the system. The equation for this control ratio is

$$Control\ ratio = \frac{G(s)}{1 + G(s)H(s)} \qquad (1.16)$$

where $G(s)H(s)$ is the open-loop transfer function. In the root locus method, the roots of the characteristic equation are plotted as a function of gain (Ref. 1.1).

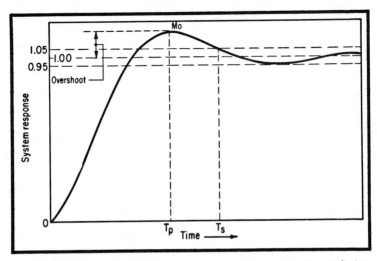

Fig. 1-6. This curve depicts the characteristic transient response to a unit step function input of a feedback loop having a single dominant pole pair. Performance parameters shown are defined in the text.

The pole-zero plot permits the selection of the most suitable pole-zero locations to obtain a specified closed-loop transfer function or transient response. The criteria for determining which poles and zeros are best must come from the specifications of system performance.

System gain is usually adjusted so that the response to a step input has the form shown in Fig. 1-6. The following figures of merit may be used to judge system performance.

• M_o, peak overshoot, is the amplitude of the first overshoot.

• T_p, peak time, is the time required to reach peak overshoot.

• T_s, settling time, is the time when the response first reaches and thereafter remains within 2 percent of its final value.

• N is the number of oscillations in the response up to the settling time.

The necessary closed loop transfer function for the time response, as shown in Fig. 1-6, may be represented by the pole-zero pattern of Fig. 1-7. The system must be dominated by one pair of complex poles and must also have the following characteristics:

• The other poles must be far to the left of the dominant poles,

28

so that the transients due to them are small in amplitude and die out rapidly.

• Any other pole which is not far to the left of the dominant complex poles must be near a zero so that the magnitude of the transient term due to it is small.

Transient response performance factors can be calculated from the pole-zero pattern of Fig. 1-7. The time required to reach peak overshoot T_p, can be expressed as:

$$T_p = \frac{1}{\omega_d}\left[\frac{\pi}{2} - \Sigma \underline{/(p_d - z)} + \Sigma \underline{/(p_d - p_n)}\right] \quad (1.17)$$

where ω_d is equal to the imaginary part of the two dominant complex poles, and represents the damped natural frequency. In the angle summation terms, p_d represents either one of the dominant pole pair, p_n represents any other pole, and z represents a zero—of the closed-loop transfer function (CLTF). The two summation terms are defined as follows:

$\Sigma \underline{/(p_d - z)}$ = sum of the angles from all zeros of CLTF to one of the dominant poles

$\Sigma \underline{/(p_d - p_n)}$ = sum of the angles from all other poles of CLTF to one of the dominant poles (the angle from its conjugate should be included in the sum).

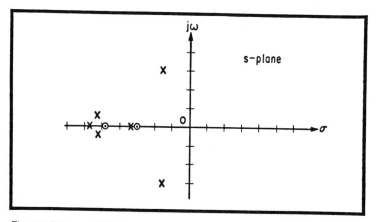

Fig. 1-7. This graph illustrates a typical pole-zero pattern which corresponds to the transient response shown in Fig. 1-6. All of the poles (X) of a stable system lie in the region to the left of the imaginary axis.

The first overshoot peak, M_0, is a function of four vector products II:

$$M_0 = \left[\frac{II\,(p_n)\,II\,(p_d + z)}{II\,(p_d + p_n)\,II\,(z)} \; e^{\sigma}T_p \right] \qquad (1.18)$$

where σ is equal to the real part of the two dominant complex poles, and overshoot time T_p is calculated from Equation 1.17. The four products are defined as follows:

$II(p_n) = $ product of the distance from all other poles of CLTF to origin (exlcuding the dominant poles)

$II(p_d - z) = $ product of distances from all zeros of CLTF to one of the dominant poles

$II(p_d - p_n) = $ product of distances from all other poles of CLTF to one of the dominant poles (excluding the distance between the dominant pole and its conjugate)

$II(z) = $ product of the distances from all zeros of CLTF to origin.

The values of T_s and N can be obtained approximately from the dominant roots. Within a 2 percent error, the settling time is equal to four time constants:

$$T_s = \frac{4}{\xi \omega_n} \qquad (1.19)$$

where ω_n is the undamped natural frequency and ξ is the damping ratio ($\omega_d = \omega_n \sqrt{1-\xi^2}$). N is the settling time divided by the period of oscillation:

$$N = \frac{T_s}{2\pi/\omega_d} = \frac{2\omega_d}{\pi \xi \omega_n} = \frac{2\sqrt{1-\xi^2}}{\pi \xi} \qquad (1.20)$$

These equations are basically valid for a system having a pair of dominant poles with all other poles far to the left of them. It is not uncommon in a process control system to have such a pole pair contributed by the dynamics of the final control element and actuator.

The calculations just outlined, when applied to the appropriate systems, can be used to examine the dynamics of control components and to improve the overall transient response. However, the controller should first be tuned according to tuning guidelines developed by Ziegler, Nichols or others as an initial step toward optimizing transient response. Controller tuning can be optimized according to RMS, ITAE, IAE or ISE criteria for any particular application. Each of these criteria offers a unique set of response characteristics to which the designer can refer in satisfying his specific requirements (Ref. 1.2).

Computer simulation techniques, either analog or digital, are better suited to analyze the stability and transient response of a complex system, or to optimize any other performance requirements than the pole-zero method just described.

SELECTING COMPENSATION HARDWARE

To successfully implement the objectives and functions required by a particular loop for optimal control, the instrument engineer must select appropriate hardware components. He should supplement his own analyses with well-established engineering guidelines developed after long experience in hardware selection. Some of these guidelines are summarized in the remainder of this article. Their application will help the control system designer to achieve satisfying results and avoid the headaches involved in redesign.

The gas pressure loop shown in Fig. 1-8 provides a basis for discussing component selection since it includes typical components required to hold output pressure constant through feedback control. The short piping run and small pressure header of this system can be modeled as a lumped parameter approximation. As was shown in Part I of this article, a first-order lag models this loop's static and dynamic

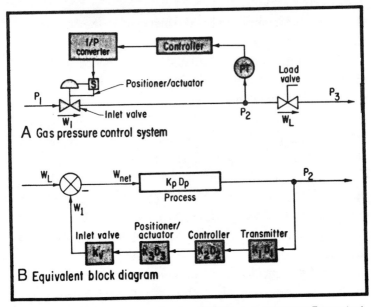

Fig. 1-8. This gas pressure control system (A) holds output pressure P2 constant through the action of components in its feedback loop. For a given range of operating conditions, the loop's static and dynamic characteristics, K and D in the equivalent block diagram (B), can be expressed adequately by a first or second-order lag.

characteristics quite adequately for a given range of operating conditions:

$$\mathcal{L}\ (P_2/W_1) = \frac{K}{\tau s + 1}$$

However, not all processes are that simple to model. In higher order systems, it is even more critical to select devices that offer operating characteristics and adjustable parameters to tailor the control scheme to the specific application.

The sensor/transmitter combination can be a source of potential problems that the control engineer needs to be aware of. These include 1) temperature sensitivity, 2) sensor rangeability, 3) linearization (for example, of a differential pressure transmitter to measure flow), 4) static accuracy and 5) vibration sensitivity. It is well worth investigating the specific requirements of each application at the initial synthesis stage in terms of these specifications and the dynamic characteristics of the loop so that a logical hardware selection can be made.

The process controller is largely responsible for the static and dynamic performance of a control loop. Its ability to function effectively depends on the availability of the ranges of its control mode settings. Furthermore, it should offer such optional functions as output limiting and anti-reset windup in addition to the standard features. This arrangement allows the instrument engineer to obtain the best performance from an individual loop.

In pH processes, where loop gain can change as much as 1,000 to 1, it is virtually impossible to configure a good control system with a standard PID controller. To achieve optimal control in this application, a nonlinear controller must be chosen.

Most process controller manufacturers offer a simple go/no-go test set which checks various circuit components; the test quickly isolates the faulty function module board. While the board is being replaced, the controller can be substituted with a manual backup unit or transferred to the manual mode.

Valve positioners can minimize overshoot and improve settling time in response to a load disturbance or setpoint change (Ref. 1.3). A positioner can improve the transient response characteristics of such relatively slow systems as liquid level, temperature, blending and some reactor control loops, where the cycling frequency is low. To compensate for the response characteristics of such fast systems as liquid pressure and flow loops, it is best to substitute a pneumatic amplifier or booster for the positioner. Positioners are required for split-range signal loops.

To stabilize a control loop, a positioner can be characterized with an input restrictor (approximately .007 to .010 in. diameter) which produces a dominant time lag (Ref. 1.4). This configuration should permit a higher proportional gain in the

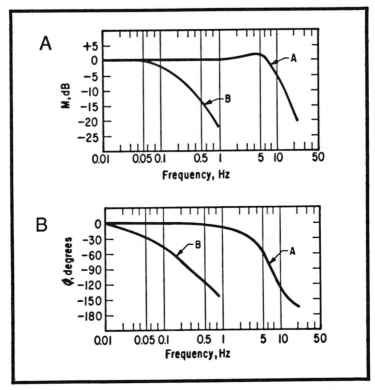

Fig. 1-9. Control loops can be stabilized with valve positioners characterized with an input restrictor which produces a dominant time lag. These plots show the frequency response of a positioner/actuator with (A) and without (B) a restrictor.

controller and, in some instances, solve a loop stability problem (if the lowest controller gain setting has already been selected). Figure 1-9 shows the frequency response of a positioner/actuator with and without an input restrictor.

Valve actuator stiffness is a function of frequency, as shown in Fig. 1-10, but this specification is commonly overlooked. In many severe service situations, large fluid forces and buffeting forces occur over a wide range of the frequency spectrum, and the actuator must be sized accordingly. The air spring rate (ASR) of a spring actuator is:

$$ASR = \frac{k\overline{P}A^2}{V} + K_a \qquad (1.21)$$

while that of a piston push-pull actuator is:

$$ASR = k\overline{P}A^2 \, (1/V_t + 1/V_b) \qquad (1.22)$$

where

$\overline{P}$ = mean cylinder (or diaphragm) pressure, psia

A = effective area (of diaphragm or piston), sq in.

V = diaphragm casing volume, cu in.

K_a = actuator spring rate, lb/in.

k = specific heat ratio

V_t = cylinder volume above piston, cu in.

V_b = cylinder volume below piston, cu in.

Control valve selection should be made with special attention to valve sizing and plug characterization. In some instances, oversizing a control valve can produce significant control problems associated with undefined gain characteristics at low lift (0 to 2 percent) and the fluid force reactions more prevalent at this lift.

Proper selection of the valve characteristic is also very important in control loop design since this parameter can help

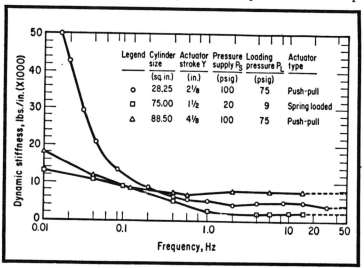

Fig. 1-10. The dynamic stiffness of a valve actuator must be considered when designing a loop that will remain stable under conditions where large fluid and buffeting forces occur over a wide frequency range. The dashed lines indicate the theoretical actuator stiffness at frequencies above 15 Hz.

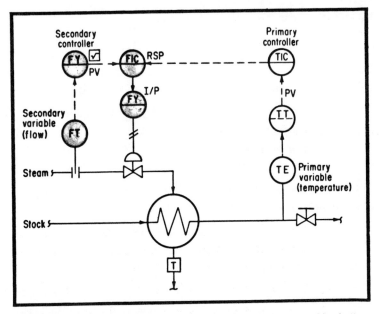

Fig. 1-11. This cascade control scheme for a heat exchanger provides better loop stability than a simple feedback system by reducing sensitivity to transients caused by load upsets.

to compensate for process nonlinearities (as shown in Part I of this article). The misapplication of a valve characteristic can severely impair loop performance and the stability of the system. Some of the other considerations in selecting the proper valve are rangeability, noise, cavitation and materials compatibility with the fluid it is modulating. All of these can be potential problem sources if the valve is not selected carefully.

IMPROVED RESPONSE WITH ADVANCED SCHEMES

Techniques other than simple feedback control, such as cascade or feedforward/feedback schemes, can further improve the transient response to a load disturbance. In the cascade loop shown in Fig. 1-11, the stock temperature setpoint is the primary controlled variable. The steam flow variations are easily compensated by the secondary control loop.

If a cascade system such as the one shown experiences instability, it may be desirable to perform the normal cascade loop startup procedures which follow. Place the primary controller in the manual mode, to see if the secondary loop cycles. If it does, the secondary controller should be retuned with a reasonable gain margin. The secondary controller generally has either P or PI control, the selection of which depends largely on the stability and control accuracy required by the inner loop. Any error set up by the secondary controlled variable is corrected through the outer feedback path of the primary controller. The primary loop which can be three-mode (PID), should be tuned by following any suitable control mode selection and adjustment procedure as if it were a simple feedback control system.

The inner loop should always be made faster than the outer loop to fully realize the transient response benefits of a cascade system. If the inner loop is slower than the outer loop, the primary controller may saturate at its output limit when a large sustained error occurs between its setpoint and the primary controlled variable.

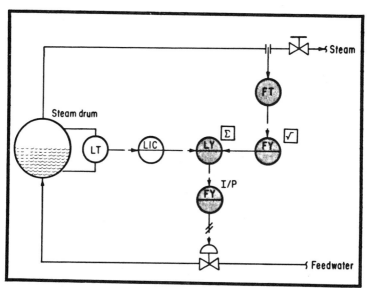

Fig. 1-12. This boiler feedwater control system, representing a feedforward/ feedback scheme, maintains the drum level at setpoint under changes in stream demand. This configuration also offers transient response characteristics that could not be obtained through simple feedback control.

The boiler feedwater control system shown in Fig. 1-12 represents a feedforward/feedback scheme, commonly referred to as two-element control in the power industry. It maintains the drum level at setpoint under variations in steam demand. In this system, the linearized steam flow measurement feeds to an adder to increase or decrease the feedwater rate. If there is any error in the level because of some mismatch between the feedforward static gain and that of the steam drum, the level feedback loop trims the level setpoint error.

The level feedback loop, containing a two-mode controller, typically cycles over a small amplitude range and at a low frequency. Cycling occurs primarily because the process transfer function is approximated as $1/As$ (integrator), with the PI modes contributing another $-90°$ phase shift; as a result, a $-180°$ phase shift exists at low frequencies.

Excluding the component failure problems which may arise in any control system, the instability of a two-element control system can be examined by first checking for stability in the level control loop. The steam flow feedforward signal should then be checked to see if it is in fact a clean signal to the adder.

REFERENCES

1.1 D'Azzo, J.J. and Houpis, C.H., *Feedback Control System Analysis and Synthesis*, McGraw-Hill Book Co., New York, 1960.

1.2 Harriott, P., *Process Control*, McGraw-Hill Book Co., New York, 1964.

1,3 Lloyd, S.G., "Guidelines for the Application of Valve Positioners," Texas A&M Instrumentation Symposium, January 1969.

1.4 Schuder, C.B., "Control Valve Rangeability and the Use of Valve Positioners," ISA/71 Conference, Chicago.

CHUN H. CHO is an applications specialist at Fisher Controls Co., Marshalltown, IA. Article is based on a paper presented at the ISA/75 Conference, Milwaukee.

Dynamic Design
of Pneumatic Control Loops

P.S. BUCKLEY

PART I: ANALYSIS BY SEGMENTS

To design any control system quantitatively, it is necessary to have 1) a design philosophy, 2) performance criteria, 3) design methods and 4) experimental data.

The writer's basic design philosophy for process control is described elsewhere (Ref 1.5, 1.6, 1.7). The feedback loop design methods on which this paper is based are the classical ones employing small signal analysis: pertubation techniques, signal flow block diagrams, Laplace transformation, and frequency response. Experimental frequency response data for transmitters, controllers, and control valves are available from various sources (Ref. 1.8, 1.9).

Different applications require different performance criteria. For example, flow control of a distillation column feed is usually far less critical than feed flow control to a high-speed reactor with a one-second holdup. In general, the control system design engineer looks for the cheapest design which is just fast enough and just sensitive enough to meet the performance requirements.

There are two basic loop configurations which are of greatest interest to the control system designer:

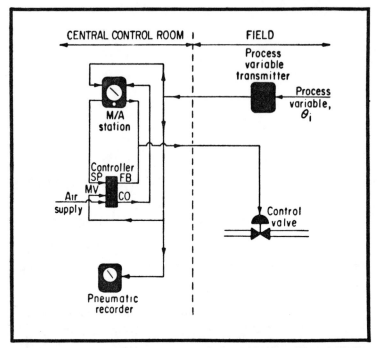

CENTRAL CONTROL ROOM | FIELD

Process variable transmitter

Process variable, θ_i

M/A station

Controller
SP   FB

Air supply   MV   CO

Control valve

Pneumatic recorder

Fig. 1-13. Two-pole configuration of a pneumatic control loop places both the controller and the manual/automatic station in the central control room. In this and following figures, signals are labelled SP for setpoint, MV measured variable, FB feedback and CO controller output.

Controller located in central control room (CCR)—This is a so-called 2-pipe or 2-tube system. Figure 1-13 probably represents 98 or 99 percent of those applications where the control station is in the CCR. It is also the basic configuration with which most of this article is concerned. It is preferred by designers because of the cost savings in comparison with 4-pipe systems.

Controllers located in the field—If the control station is in the CCR but the controller is in the field, four pipes or tubes are required to go to the field as shown by Fig. 1-14. This design was popular in the early 1950's before the work of Catheron (Ref. 1.10) became well-known. This work showed that the pneumatic transmission lags of 2-pipe systems could be greatly reduced by terminating the controller output line with a booster relay or valve positioner.

In most loops with field-located controllers, the valve, transmitter, and controller are within 10 to 20 ft of one another. In all cases, transmission lag can be minimized by inserting isolation relays (1:1 pressure repeaters) in branched lines coming back to the CCR, Fig. 1-15.

DYNAMIC EFFECTS OF LOADING

It has been recognized for some years that the dynamic output/input ratio (transfer function) of a device is very much affected by the load. The frequency response of a ΔP transmitter with a load of 10 ft of ¼-in. OD dead-ended tubing is vastly different from that of the same transmitter with 2000 ft of ⅜-in. OD dead-ended tubing. The ISA RP-26 committee recognized this problem and developed a set of standardized test procedures (Ref. 1.11), intended for use by the designers

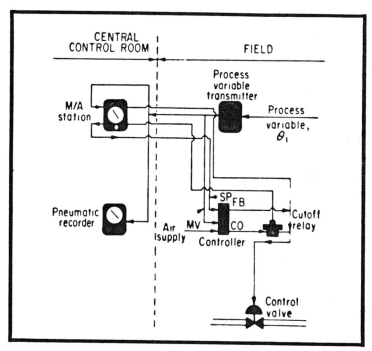

Fig. 1-14. Four-pipe configuration locates the controller in the field and requires additional field transmission lines to link the controller and the panel-mounted manual/automatic station.

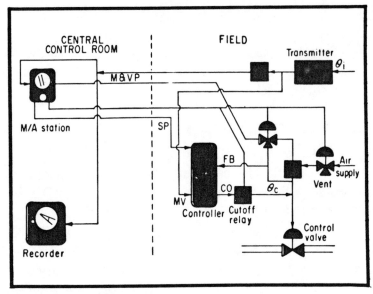

Fig. 1-15. Transmission lags can be minimized in a four-pipe system by inserting pressure repeaters or isolation relays (marked 1:1) in the lines.

of pneumatic devices as well as by those who design pneumatic control systems.

The ISA procedure calls for making measurements at the input and output of each device and at the input and output of the transmission line as shown in Fig. 1-16. For example, to determine the performance of a device with a ¼-in. OD transmission line 200 ft or more in length, and with a small terminal volume requires:

- Output/input frequency response tests with the device loaded by 200 ft of ¼-in. tubing terminated in a pneumatic receiver element having a volume of approximately 1.2 cubic inches
- Output/input frequency response tests on the transmission line of interest
- Preparing Bode plots for each of the two tests
- Adding the magnitude and phase angle curves to prepare a third, composite Bode plot of the device and its load.

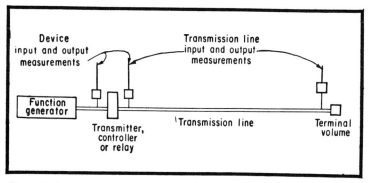

Fig. 1-16. Fig. 1-16. Test configuration to determine frequency response of a pneumatic device and transmission line in accordance with ISA standard S26.

The designer of pneumatic *devices* can make effective use of this detailed test procedure, although the standard is deficient in one important respect; it makes no provision for testing a device with a ⅜-in. tubing load terminated with a small volume. But the designer of pneumatic control *systems* would like to have a simpler procedure to determine the frequency response from the input of a device to the far end of the output tubing.

To avoid interaction between the device and its load, we have adopted a test procedure which divides the pneumatic control loop into a number of non-interacting segments. Frequency response measurements are made from the input of one device to the input of the next device; or more generally, from the input of one device to the end of its output line, which is assumed to be terminated in a small volume. For the systems designer, this procedure has the following advantages over that of ISA-S26:

- Testing is simpler, somewhat more accurate, and ⅜-in. OD lines with small volume terminations are readily accommodated
- For a given device and load, only one Bode plot is required, instead of two or three
- The signal flow block diagram for the control loop contains fewer elements.

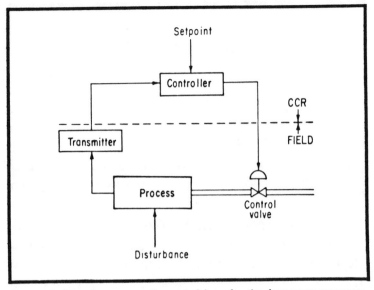

Fig. 1-17. Two-pipe pneumatic control loop for the frequency response analysis-by-segments discussed in text.

Test data obtained by this procedure for various instruments with output transmission lines up to 250 ft are given in Reference 1.8. For lines ranging in length from 500 ft to 2,000 ft, both ¼-in. and ⅜-in. OD plastic tubing, data are presented in Reference 1.9.

LOOP SEGMENT ANALYSIS

The validity of the segmental test is supported by extensive mathematical studies which indicate that as far as load impedance effects are concerned the vast majority of pneumatic controllers, transmitters, computer relays, etc. can be analyzed by a "general transmitter theory" (Ref. 1.8, 1.12). Under this procedure, the complete design by frequency response methods of the 2-pipe control loop shown in Fig. 1-17 involves obtaining the frequency response of four segments.

Transmitter input to controller input—For such devices as pressure and differential pressure transmitters which are close-coupled to the process, testing is straightforward. For such devices as temperature and displacer-type level transmitters, testing is more difficult (Ref. 1.13, 1.14), and in the case of temperature it is impractical to run tests for all possible heat transfer coefficients. But typical temperature transmitter dynamics are very similar to pressure or differential pressure transmitter dynamics, except for lags in the thermal detector. For a bare bulb, one may readily calculate a first-order lag; while for a bulb in a protecting well, a second-order lag may be calculated (see Chapter 22 of Ref. 1.5). The temperature transmitter's frequency response is therefore equal to that of a pressure or differential pressure transmitter multiplied by the appropriate bulb lag.

For displacer-type, force-balance level transmitters the same comments apply. One may use the frequency response of a pressure or differential pressure transmitter multiplied by the second-order displacer dynamics (see Chapter 18 of Ref. 1.5). For other types of displacer transmitters, direct testing is advisable.

It's interesting to note that for very long transmission lines, all pressure or differential pressure transmitters—regardless of make—have almost the same frequency response for a given line length and diameter. For very short lines—say 10 to 30 ft—the differences are more marked. This is where the ISA-S26 tests are particularly useful to the equipment designer.

The long line tests reported in Reference 1.8 were run in 1965 and 1966 on a Foxboro 13A d/p cell with a 1.0 scfm pilot (nominal capacity). These tests showed that for a given line length (over a range of 500 to 2,000 ft):

- ⅜-in. OD plastic tubing (0.250-in. ID) provides significantly faster transmission than does ¼-in. OD plastic tubing (0.170-in. ID). The use of ⅜-in. OD plastic tubing permits doubling the speed of response of a properly designed liquid flow control system for one-way transmission distances of 500 ft or more.
- Transmitter pilot capacity somewhat limits frequency response, as can be seen by comparing the response

of transmitter plus line with that of the line only. This is true for both ¼-in. OD and ⅜-in. OD tubing.

In the last couple of years, Foxboro has increased its d/p cell pilot capacity to a nominal 2 scfm, which should bring the performance of this device plus output line even closer to that of the line only.

Incidentally, a 1,000-ft line has more lag than two 500-ft lines connected by an isolating relay. This means that the significant figure to use for transmission distance is that from the output of one device to the input of another. In most cases, the control valve and transmitter are fairly close to each other in the field; the one-way transmission distance is therefore the significant figure. This practice of rating transmission distance is even more important in designing complex controls; for example with overrides, where the transmission distance from transmitter to the CCR may be quite different from that of the CCR to the valve.

According to the "general transmitter theory" mentioned earlier, one may think of a pneumatic force-balance transmitter as a proportional-only pressure controller. Its setpoint is the process variable, and calibration consists of adjusting controller gain and zero such that the 3 psig output corresponds to a particular value of process variable and a 15-psig output corresponds to another. The pressure transmitter (controller) is so designed that its *immediate* output pressure is the pressure which is controlled. Unfortunately, what is really desired is a design which controls the pressure at the *far end* of the transmission line as quickly as possible. Although transmitter designs to accomplish this have been studied (Ref. 1.15) and the necessary information is available, no commercial devices incorporate the necessary circuitry.

Controller input to valve input—Much the same comments may be made in connection with this loop segment as were made for the transmitter-controller segment except for the addition of three cautions:

- The test configuration should include the manual/automatic switching station
- The output end of the line should be terminated in a small volume—such as a booster relay or positioner
- If the control station does not have a built-in booster on its output side, one should be installed as shown in Fig. 1-18.

Controllers connected with 200 ft of line to a small volume termination have shown slightly faster frequency response than did transmitters with the same load. The reason for this is discussed later (under *"Proportional band adjustment"*). To minimize the amount of testing, we no longer perform extensive tests on controllers. Instead, for design purposes we use transmitter data obtained with lines 500 ft or longer.

Valve input to valve stem position—Controller-to-valve lags can be minimized by equipping the valve with either a

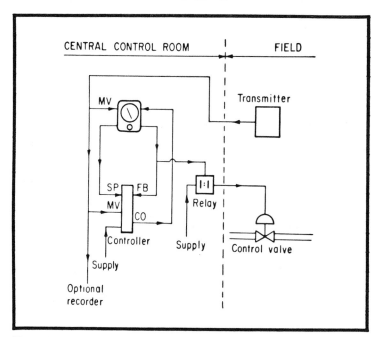

Fig. 1-18. Installation details for booster relay on control station output.

booster or valve positioner as noted above. Valve positioners are preferred for the following reasons (Ref. 1.16):

1. In valves without positioners, stiction and dead spot effects are large enough (typically 0.25-0.50 psi) to make frequency response design methods invalid. With a positioner, dead spot is usually reduced by a factor of 10 or more. In terms of positioner input signal, dead spot is commonly in the range of 0.012 to 0.050 psi. (It is possible, by computer simulation, to design a control loop when significant stiction effects are present, but only if the magnitude and nature of these effects are known. Since it is impossible to predict in advance how tightly a mechanic will compress a valve stem packing, simulation is not usually useful to design a loop whose valve has no positioner.)

2. With a positioner, the valve's inherent flow characteristics are maintained. Without a positioner they are not—due to stiction and coercive, axial stem forces.

3. A positioner-operated valve permits much closer control than can be achieved by a valve without a positioner.

Most manufacturers can now furnish frequency response data for their valves. Generally this data will apply to the combinations of valve plus positioner. Due to the loading effect of the valve topworks plus the influence of valve packing friction, it is not feasible to run frequency response tests separately on the positioner and valve. For some of the newer designs where positioner gain is in the range of 500 to 1,000, separate testing would be even more difficult.

Valves equipped with "failsafe" springs usually show poorer dynamic performance than valves without such springs. The stiffer the spring, the poorer are both the frequency response and the resolution. Furthermore, the stiffer the spring, the less ability the valve has to offset large, axial stem forces. For valves larger than 6-in. and with a high pressure drop, it is sometimes hard to find a suitable actuator.

Valve stem position to process variable—Normally this response is calculated from process design details, rather than being measured. Methods of prediction are given in References 1.5 and 1.17. For applications such as most flow controls, flow ratio controls and liquid level controls, simple formulas and charts are available, or can readily be prepared, which reduce the time required for design calculations to a few minutes. Where it is desired to measure experimentally the relationship between valve stem position and process variable, one may use procedures such as those discussed by Hougen in Reference 1.18.

A signal flow block diagram for the control loop of Fig. 1-17 is shown in Fig. 1-19, with terminology as defined in the Table 1-1. Note that it is sometimes more convenient to express $K_vG_v(s)$ in terms of flow change per change in valve positioner input signal. In this case $K_vG_p(s)$ is expressed as process variable change per change in manipulated flow.

Having obtained the frequency response data for each segment of the control loop, the engineer may proceed to determine controller gain, reset, and derivative times in addi-

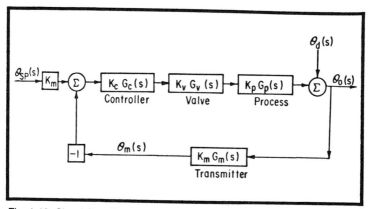

Fig. 1-19. Signal flow diagram for the control loop of Fig. 1-17. Symbols are defined in Table 1-1.

Table 1-1. Signal Flow Terminology

$\theta_{sp}(s)$	= controller setpoint in terms of process variable
K_m	= process variable transmitter static gain, $(\Delta\theta_m)_{max}/(\Delta\theta_o)_{max}$
$(\Delta\theta_m)_{max}$	= transmitter output span, psi (usually 12 psi)
$(\Delta\theta_o)_{max}$	= transmitter input span corresponding to $(\Delta\theta_m)_{max}$
$K_r G_r(s)$	= transfer function relating change in valve positioner input signal to change in controller input signal
$K_t G_t(s)$	= transfer function relating change in valve stem position to change in valve positioner input signal
K_t	= gain of the valve, or ratio of inches stem travel per psi to valve positioner input
$K_p G_p(s)$	= transfer function relating change in process variable to change in valve stem position
$K_m G_m(s)$	= transfer function relating change in signal at end of transmitter transmission line to change in process variable
$\theta_d(s)$	= disturbance
s	= Laplace transform variable (substitution of $j\omega$ for s converts the expression to frequency response)

tion to closed-loop resonant frequency. Since so many commonly-encountered systems have been extensively studied, it's often necessary only to look up the pertinent literature.

FACTORS AFFECTING LOOP PERFORMANCE

The dynamics of pneumatic control loops are limited not so much by the intrinsic properties of pneumatics as by the conventions of (a) instrument application and (b) instrument design (Ref. 1.9). The latter are largely beyond the control of the system designer so the following discussion is limited to application conventions. Internal diameter of tubing, transmission line termination volume, presence or absence of restrictions in manual/automatic control stations, adequacy of field air supplies, controller proportional band adjustment, and field installation practices for transmitters and valves are all factors in loop dynamics.

Tubing diameter—As noted earlier, ⅜-in. OD tubing, whether metallic or plastic, provides much faster response than does ¼-in. OD tubing. Because of the much lower installed cost, plastic tubing is now widely used.

Termination volume—With rare exceptions, it is customary to terminate pneumatic transmission lines in a volume such as a receiving bellows, diaphragm chamber, or valve topworks. Lines terminated in a small volume respond faster than those terminated in a large volume. In our early tests, we frequently terminated the lines with two 3-½-in. Bourdon receiver gauges in parallel. But especially for long lines, it seems to make little difference whether the line is dead-ended, has one receiver gauge, or two.

With very short lines, however, some pneumatic devices are unstable when dead-ended or connected to small volumes. Where pneumatic devices are close-coupled it has sometimes been necessary to insert volume pots. This is fortunately much less of a problem today than 20 years ago.

Any optimum termination is physically and mathematically much more complex than a simple volume. For the smoothest rolloff in the frequency response amplitude curve (i.e., a rolloff without resonances) and for minimum phase shift at a given amplitude ratio, the line should be terminated in what is called its "characteristic impedance" (Ref. 1.5, 1.8). Although the selection of proper termination impedance has been an enormously useful concept in electrical engineering, it unfortunately gets little attention in pneumatics.

Restrictions in control station—Some years ago, we were concerned about possible limitations on controller performance due to restrictions in the control station manual/automatic switch block. To test this, a popular controller and station were piped up (plug-in manifold was not used) and loaded with 250 ft of ⅜-in. OD copper tubing terminated in the topworks of a 4-in. valve with a spring-and-diaphragm ac-

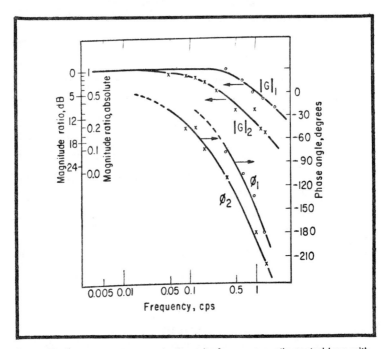

Fig. 1-20. Frequency response test results for a pneumatic control loop with (Case 2) and without (Case 1) a manual/automatic control station.

tuator. Output pressure was measured at the topworks and a frequency response test was run. A second test was performed, for which the control station was removed and the controller output connected directly to the 250-ft line. The results, plotted in Fig. 1.20, show that switch block restrictions can have a devastating effect on frequency response; much greater, in fact, than the effect of changing from ⅜-in. to ¼-in. tubing. Theoretically, the effect should be somewhat less if the ⅜-in. OD line were terminated in a small volume and considerably less for a ¼-in. OD line terminated in a small volume.

The combined effects of switch block and plug-in manifold restrictions probably explain why Warnock's tests (Ref. 1.19) failed to show more than a small benefit from using ⅜-in. OD plastic tubing instead of ¼-in. OD tubing in a liquid flow control loop.

The insertion of a 1:1 booster relay on the output side of the control station (see Fig. 1-18) compensates for the loss in performance caused by the station restrictions. This combination has been effectively used with ⅜-in. OD tubing in a large number of applications where the one-way transmission distance is 500 to 1,000 ft.

For one-way transmission distances of 250 ft or less, ¼-in. OD tubing can be used without 1:1 relays. The test data for such runs, as indicated earlier, include the effect of control station restrictions.

Some vendors build a booster into their control station on the output side of the switch block. Others use what is in effect a 4-pipe system, with the controller located in the central control room and the controller output going through a cutoff relay instead of a switch block.

Field air supplies—The adequacy of the air supply is worth far more consideration than can be allocated to it here. In distillation column control systems, air supply fluctuations of 1-2 psi on flow and pressure transmitters and 2-5 psi on level transmitters are frequently seen. Some difficult control problems have vanished when cheap air supply regulators were replaced by first-class, precision supply regulators.

Commonly-encountered shortcomings in field air supplies include:

- Poor quality regulator, with an excessively steep pressure-vs-output flow curve
- A supply line between regulator and transmitter whose length is too great and whose internal diameter is too small
- Too many transmitters supplied by the same regulator
- Transmitters and valve positioners supplied by the same regulator.

For most applications, although not all, a ⅜-in. OD or larger supply line no more than 10 ft long is recommended from the regulator to each transmitter. A regulator should be selected whose droop between 0.5 scfm and $2n$ scfm is no more than 0.05 psi—where n is the number of load (transmitters, controllers, and relays).

Valve positioners should have an air supply separate from that of transmitters. Most modern positioners can operate directly off 60-150 psi air headers and will perform much better at these pressure levels than at 20-25 psig. Because the newer 2-stage designs have smaller diameter restrictions in the input pilot (than do older single-stage positioners), each positioner of this type should be provided with its own filter (not filter-regulator).

Transmitters should have test connections at the air supply ports so that the adequacy of field air supplies can be verified. Small gages of the 1- to 2-in. diameter variety, which are frequently supplied with inexpensive regulators, have very poor accuracy. Tests revealed that many are in error by 2 to 6 psig, and one was found which read 11 psig high. If supply pressure is fluctuating significantly, the addition of a volume pot (say 1 to 2 gallons capacity) at the supply port will usually dampen the fluctuations.

Proportional band adjustment—In a mathematical model of a controller with a proportional circuit of the pressure-dividing type, the equations predict faster response for a low gain setting (wide proportional band) than for a high gain setting. Experimental work confirms this. The effect is not large, however, until one reaches a gain of 10 or higher (10 percent or narrower PB). Since only a few applications require such high gain, this phenomenon has been generally disregarded. But for any new pneumatic controller design, the vendor should be asked if gain setting will have a significant effect on frequency response.

For tubing runs up to 250 ft, a controller with a proportional band of 100 percent or greater will be faster than a transmitter with the same load. We have no test data for proportional band effect at distances of 500 ft and more, but would expect (from theoretical considerations) that at least qualitatively the effect would be the same as that observed at 250 ft.

For the majority of applications with one-way transmission distances of 500 ft or more, we have used transmitter data as valid response data for the transmitter or for the controller. This is a conservative design procedure because the controller is expected to be a little faster.

Transmitter installation—This is another important topic which warrants more discussion than can be given here. Lags and other measurement problems are often due more to installation practices than to hardware limitations. Problems with liquid level measurements associated with distillation columns have been reviewed in Reference 1.20. Shortcomings were also found in typical temperature measurements. For example, in the control of condenser subcooling (by throttling the cooling water) one can improve the speed of response of the temperature measurement (as much as 10- to 100-fold) by funnelling condensate over a bare bulb immediately beneath the condenser, instead of using a bulb in a well (frequently not tight fitting) located in the reflux drum. The latter technique almost always introduces a measurement lag which is much greater than that of the condenser, so substantial control improvement can result from installing a bare bulb below the condenser.

Valve positioners—The adverse effect of control valve lags on control loop performance has often been greviously underestimated. A (small) 1-in. valve equipped with a piston actuator without a failsafe spring and with a modern positioner (2-stage, double-acting, high capacity) will be 10 to 50 times faster than a (large) 12-in. valve with a spring-and-diaphragm actuator and an inexpensive single-stage positioner.

With this kind of performance variation, it is difficult to generalize about the relative magnitudes of valve and transmission line lags. Nevertheless, it appears that the fastest available valves may contribute more lag than the transmission lines for one-way distances less than 100 ft, and that slower valves contribute the major lag for distances up to 1,000-1,5000 ft—even when ⅜-in. OD tubing is used.

Valves up to the 2- or 3-in size are usually fast enough even when equipped with a positioner of average speed. The performance data for 4- and 6-in. valves should be scrutinized with care. For larger valves, only piston actuators with double-acting, high capacity positioners and without fail-safe springs should be considered. Available actuator thrust (or torque) should be at least double the maximum which the process can exert on the stem.

REFERENCES

1.5 Buckley, P.S., *Techniques of Process Control,* John Wiley, New York, 1964.

1.6 Buckley, P.S. *Chemical Engineering Progress,* May 1969, pp. 45-51.

1.7 Buckley, P.S., "A Modern Perspective on Controller Tuning," Texas A&M 30th Annual Symposium on Instrumentation for the Process Industries, January 1973.

1.8 Grabbe, E.M., *Handbook of Automation, Computation, and Control,* Vol. 3, Chapter 7 by P.S. Buckley and J. M. Mozley, John Wiley, New York, 1961.

1.9 Buckley, P.S. and Luyben, W.L., "Designing Long-Line Pneumatic Control Systems," *Instrumentation Technology*, April 1969, pp. 61-66.

1.10 Catheron, A.R., "Factors in Precise Control of Liquid Flow," Paper 50-8-2, *Proceedings of the Fifth ISA National Instrument Conference and Exhibit*, September 1950, Buffalo.

1.11 ISA-S26 Standard, "Dynamic Response Testing of Process Control Instrumentation," 1968.

1.12 Buckley, P.S., "Dynamics of Pneumatic Control Systems," Paper 55-6-2, *Proceedings of the 10th Annual ISA Instrument-Automation Conference and Exhibit*, September 1955, Los Angeles.

1.13 Looney, R., "A Thermal Sine Wave Generator for Speed of Response Studies," ASME Paper 54-5A-38, 1954.

1.14 Higgins, S.P., "A Thermal Sine Wave Apparatus for Testing Industrial Thermometers, ASME Paper 54-5A-20, 1954.

1.15 Vannah, W.E. and Catheron, A.R., "Improved Flow Control with Long Lines," Paper 51-6-2, *Proceedings of the Sixth National Instrument Conference and Exhibit,* September 1951.

1.16 Buckley, P.S., "A Control Engineer Looks at Valves," *Proceedings of the First ISA Final Control Elements Symposium*, May 1970, Wilmington.

1.17 Campbell, D.P., *Process Dynamics*, John Wiley, New York, 1958.

1.18 Hougen, J.O., *Measurements and Control Applications for Practicing Engineers*, Cahners Books, Boston, 1972.

1.19 Warnock, J.D., "How Pneumatic Tubing Size Influences Controllability," *Instrumentation Technology*, February 1967, pp. 37-43.

1.20 Buckley, P.S., "Liquid Level Measurement in Distillation Columns," *ISA Transactions*, Vol. 12, No. 1, 1973, pp. 45-55.

PART II: APPLICATION PRINCIPLES AND PRACTICES

Control loop design techniques are necessarily contingent on an overall process control design philosophy (Ref. 1.21, 1.22). Most controls can be divided into three categories:

- Material balance controls
- Product quality controls
- Constraint or protective controls

The most common material balance controls are the familiar averaging liquid level controls. These function as filters for flow disturbances which might otherwise upset the product quality controls. The slower the level controls are, the better filtering they provide. By including this filtering action or feedforward compensation (or both) in the overall control scheme, one can greatly minimize the need for speedy response in the product quality controls.

An appropriate control loop design therefore should achieve a dynamic balance between the three categories of controls. To provide adequate overall control performance, the designer should recommend holdup sizes for level control, in addition to calculating and specifying all controller settings for all categories of controls in advance. Calculations cannot be carried out, however, until two questions have been answered for each loop: How fast is fast enough? How precise is precise enough?

HOW FAST IS FAST ENOUGH?

One can begin by trying to estimate the closed-loop resonant frequencies of product quality control loops if standard commercial transmitters, PI controllers and valves are used, but without considering pneumatic transmission lags. Then one can design averaging level controls to lower the closed-loop resonant frequencies by a factor of 10. Usually, there is little benefit in going beyond a factor of 10. If required holdups are too large, one can try to substitute feedforward compensation; but doing the job with holdups has the advantage of reducing the required flow turndown.

Once adequate holdup and/or feedforward compensation is provided, the effect of pneumatic transmission lags can be checked next. As a performance criterion, it can be arbitrarily decided that 2-pipe pneumatic transmission will be satisfactory if it does not introduce more than 20-degree additional phase lag at the resonant frequency of the closed-loop system f_n (where f_n is determined without considering the pneumatic transmission lag). Tables to facilitate these calculations are given in Reference 1.23.

In controlling the top temperature of a distillation column by manipulating reflux flow, one can assume as an example that the closed-loop resonant frequency f_n without pneumatic transmission lags has been calculated as one cycle in 6 min. From the relationship $f_n = 0.16/\tau_2$, τ_2 is found as 0.95 min and, from Table 1-2, one could therefore use 2,020 ft of ⅜-in. OD plastic tubing each way in the loop. For a smaller value of f_n, i.e., a slower temperature control, one could use an even longer run of ⅜-in. tubing.

Assuming that in the same distillation column the top temperature control is cascaded to reflux flow control or to reflux-to-feed flow ratio control, a new factor needs to be considered. For cascade control to be worthwhile, the secondary or slave control loop should be at least five (preferably 10) times faster than the master loop. If the flow or flow ratio loop is to have a closed-loop resonant frequency which is 10 times faster than the master loop, then its frequency is 10 cycles in 6 min, or 1.67 cpm; for a five-times-faster loop, the correspond-

Table 1-2. Resonant Frequencies for Level Loops

Tubing length (ft)	Natural frequency f_n, cpm 1/4 in. OD	Natural frequency f_n, cpm 3/8 in. OD
250[1] copper	1.5	
250[2] copper	9.0	
250[3] copper	22.8	
505 plastic	4.5	9.0
1,010 plastic	1.5	3.5
1,515 plastic	0.72	1.4
2,020 plastic	0.39	0.78

Notes: 1. ID = 0.18 in.
2. Includes restrictions in automatic-manual station.
3. 4-pipe system, field-located controller.

ing frequency is 0.84 cpm. According to Table 1-2, 0.78 cpm can be achieved with 2,020 ft of ⅜-in. OD tubing—which would satisfy the five-times-faster cascade loop. But for the 10-times-faster loop, the run of ⅜-in. OD tubing is limited to about 1,515 ft, for an f_n of 1.4 cpm.

Little advantage is achieved by making slave loops much faster than 10 times the master loop. For a single loop, there is correspondingly little advantage in making the transmitter or control valve more than 10 times faster than the closed-loop resonant frequency.

Closed-loop resonant frequency is regarded as one of the critical factors in process control. The proper selection of this frequency for each loop is one of the major methods of eliminating or avoiding interaction. Closed-loop frequency response can be quickly calculated by Nichols plots directly from open-loop frequency response curves (Ref. 1.21), Chapter 8). All other representations of dynamic behavior, such as step or pulse response, must be extensively reprocessed, often by a computer, to convert them into frequency response form.

HOW PRECISE IS PRECISE ENOUGH?

It is difficult to set up experiments to test effects of sticking and dead spots. Therefore, computer simulation is preferred. These studies have shown that dead spot, dead zone, or threshold sensitivity should be 10 percent or less of the desired control range. If, for example, it were known that the distillation column temperature control system mentioned earlier would be adequate if temperature were held within ±1 °C, then the temperature measurement dead spot should be no larger than ±0.1 °C. This is not difficult to achieve since a typical transmitter with a 50 °C span will have a resolution of about 0.05 °C.

Satisfying this criterion does not in itself guarantee good control, and greater resolution (smaller dead spot) improves control by very little. It is difficult to tell the difference between systems with no dead spot and those with dead spots that are less than 10 percent of the desired control range.

RECOMMENDED LOOP DESIGN PRACTICES

There is no way to guarantee satisfactory performance of a control system without designing it quantitatively. Nevertheless, there are some general design practices which are useful.

In most cases, valve speed is not important. Slow valves should be avoided, but a high performance control valve is only necessary in certain critical 4-pipe systems. All valves should be positioner-operated.

An adequate air supply is needed for all transmitters, valve positioners and field-located controllers. Design guidelines for air supply are:

- Use no more than 10 ft of ⅜-in. OD tubing from the regulator to each load
- Use a separate, high pressure (60 psig or greater) air supply for positioners
- Use a filter at the positioner supply port
- Supply line length and diameter should be tailored to the requirements of each positioner
- Provide a test connection at the supply port of each field-located device and at outputs of transmitters, controllers and relays.

Transmission distance guidelines for 2-pipe and 4-pipe systems are shown in Table 1-3.

APPLICATION GUIDELINES

In addition to the general guidelines just cited, the dynamics of particular control system applications may call for additional precautions or special care in design. Here are some further considerations in pneumatic loop design, on a unit operations basis:

Compressor antisurge controls—A 4-pipe system is generally needed for fast response. The control valve should be

Table 1-3. Transmission Distance Design Practices

	Tubing size (OD)	Booster requirements	Frequency Response Calculations	Comments
2-pipe, up to 500 ft	1/4 in.	Use 1:1 booster on station output or a 4-pipe hookup with controller in central control room (CCR)	Use experimental data provided by each vendor for transmitter, controller plus station, and control valve components	A 4-pipe hookup eliminates problems caused by manual-automatic station restrictions because the controller output does not go through the manual-automatic station
2-pipe, 500 to 2,000 ft	3/8 in.	Use 1:1 booster after manual-automatic station if it has a restrictive switch block, or use a 4-pipe hookup with controller in CCR	Use transmitter data for transmitter and controller-station combination. Choice of vendor is immaterial	For 2-pipe systems with one-way distances of 500 ft or greater, use a transmitter and controller (or 1:1 booster) with pilot capacities of 2.0 scfm or greater
2-pipe, over 2,000 ft	3/8 in.	Same as above	Consult transmitter and controller vendors for recommendations and extrapolate the long-line performance only if actual measurements are not available	Use a "line shrinker" (Ref. 4) at the transmitter and controller outputs. The line shrinker is a device which applies 2- or 3-stage derivative compensation for transmitter line lags
4-pipe, with field-located controller	1/4 in.	Do not use booster with 4-pipe systems	Use experimental data provided by each vendor for transmitter, controller plus station, and control valve components	In critical applications, use isolation relays for transmitter and controller signals returning to the CCR. This will result in performance comparable to a field-located controller that does not transmit to the CCR

equipped with a piston operator and with a 2-stage, push-pull, high-capacity positioner; the valve should be capable of full travel in no more than 3-5 s.

For more critical applications, such as high pressure ammonia and methanol systhesis gas compressors, an even faster system may be needed. The two signal lines returning to the CCR, Fig. 1-21, should be isolated in the field by 1:1 relays. The antisurge valve should be capable of full travel in about 1 s, which might necessitate the use of a hydraulic operator.

Chemical reactor with small holdup time—Feed ratio controls or reactor temperature controls should have the same design as was described for compressor antisurge controls. The temperature measurement should be high speed, preferably a bare bulb, but possibly a small diameter sheathed thermocouple in a small, tight-fitting thermowell. A 4-pipe system is usually needed.

Pressure control in polymer headers—In the manufacture of synthetic fibers, it it usually required to accurately control

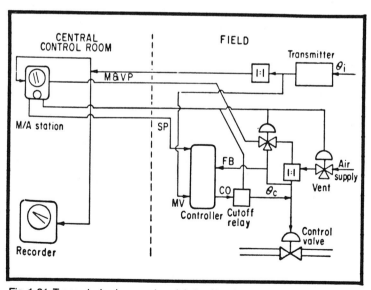

Fig. 1-21. Transmission lags can be minimized in a four-pipe system by inserting pressure repeaters in isolation relays (marked 1:1) in the lines.

the pressure of molten polymer or polymer solutions just ahead of the spinnerets. A 4-pipe system is generally needed.

Flow and flow ratio controls—Since these are almost always slave loops whose master loops are either product quality or material balance controls, the choice between 2-pipe or 4-pipe systems depends mostly on the speed of the master loop. The same comment applies to permissible transmission distance.

Temperature controls—Temperature control of gas-gas heat exchangers will normally be slow, unless a bypass control system is employed. Pneumatic transmission distance is unlikely to be the limiting lag in the system. Liquid-liquid heat exchangers are usually faster; but even so, process lags will be the limiting ones unless a bypass control scheme is used.

For very critical control applications, a bypass control scheme and a fast temperature measurement will be used on either gas-gas or liquid-liquid exchangers. Such a control scheme is inherently rapid, and in some cases a 4-pipe configuration will be required.

Averaging level controls—With typical averaging level control time constants of 15 min or more, transmission distance is almost never a problem. Associated auto overrides, however, may require more care, but transmission distances up to 2,000 ft one way will seldom cause problems.

Distillation columns—For most applications there will be no difficulty with properly designed 2-pipe systems if one-way transmission distances do not exceed 1,500-2,000 ft.

Composition controls—For large or slow reactors or for mixing and blending systems, a one-way transmission distance of 1,500-2,000 ft will usually be acceptable. Some waste treatment systems, however, may have to contend with such large and fast disturbances that even though holdup capacities are also large, control must be reasonably agile. Because such systems are often more than 2,000 ft from the CCR, 4-pipe loops can be required.

PERSPECTIVE ON LOOP DYNAMICS

In the last 25 years the author and his associates have carried out quantitative studies on thousands of control loops. Contrary to some popular opinion about the magnitude and influence of transmission lags, it has been found that:

- Transmission lags are small enough in most properly designed pneumatic loops (say 99 percent of loops other than flow and pressure) that they do not affect dynamic performance calculations.
- Where transmission lags are large enough to be included in the calculations, they rarely have a significant effect on control loop performance.

Some hundreds of loops with ⅜-in. OD tubing and one-way distances of 500-1,000 ft are operating satisfactorily. The controller settings and closed-loop resonant frequencies are about as calculated. We know of one level control loop with a one-way distance of 3,200 ft (¼-in. OD tubing) which is operating satisfactorily with a 50 percent PB (proportional-only) controller.

In most existing pneumatic loops the limitations on performance are causesd by:

- Use of control valves without positioners
- Use of ¼-in. OD rather than ⅜-in. OD tubing
- Restrictions in manual-automatic switch blocks and plug-in manifolds
- Inadequate valve positioner air handling capacity, or use of single-acting positioner (instead of double-acting) on cylinder operators
- Inadequate field air supplies
- Multiple process lags
- Industry practice of designing transmitters, controllers and relays to come to balance on the immediate output pressure rather than on the pressure at the far end of the signal line.

REFERENCES

1.21 Buckley, P.S., *Techniques of Process Control,* John Wiley, New York, 1964.

1.22 Buckley, P.S., "A Modern Perspective on Controller Tuning," Texas A&M 30th Annual Symposium on Instrumentation for the Process Industries, January 1973.

1.23 Buckley, P.S. and Luyben, W.L., "Designing Long-Line Pneumatic Control Systems," *Instrumentation Technology,* April 1969, pp. 61-66.

1.24 Vannah, W.E. and Catheron, A.R., "Improved Flow Control with Long Lines," Paper 51-6-2, *Proceedings of the Sixth National Instrument Conference and Exhibit,* September 1951.

PAGE S. BUCKLEY is a Principal Design Consultant in the Engineering Department of E.I. du Pont de Nemours & Co. at Wilmington, DE.

Batch Reactor Control

W.L. LUYBEN

Batch chemical reactors offer some of the most difficult challenges to control engineers. Since a batch process is inherently dynamic or time-varying in nature, process variables can change greatly during the course of a batch cycle. No normal steady-state operating condition exists at which controllers can be tuned for good dynamic response.

Probably the most challenging control problem presented by batch reactors is the open loop instability caused by reactions which produce heat. A small increase in temperature can make a reaction proceed more rapidly, which generates more heat and raises the temperature even higher. Runaway reactions can occur if temperatures are not controlled properly.

Other control problems include batch-to-batch variations in the chemical reaction caused by catalyst activity or reactant purity, reset windup in conventional controllers and large changes in process gains and time constants. Economic incentives for close control of batch reactors include maximizing capacity, improving batch-to-batch uniformity and product quality, and safety.

Fig. 1-22. A typical batch reactor installation.

BATCH CYCLE CONTROLS

A simple batch reactor control system consists of a temperature measuring device, a two- or three-mode controller and a control valve which throttles coolant flow, Fig. 1-23. The setpoint signal to the temperature controller may be constant, or it may be varied during the batch, according to an optimum temperature profile, in order to produce the maximum amount of product in a minimum amount of time.

At the beginning of a typical batch cycle, reactant material is charged into the reactor vessel. Typically, the reactor is heated to a higher temperature and the reaction begins to occur. If the reaction itself produces heat, temperature of material in the reactor will begin to increase to a point above the optimum process temperature, and heat must be removed to maintain a constant temperature.

Reactions which produce heat are *exothermic*; reactions which consume heat are *endothermic*. An *isothermal* reactor is

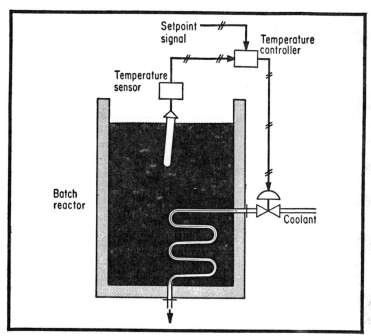

Fig. 1-23. Typical batch reactor with temperature control system.

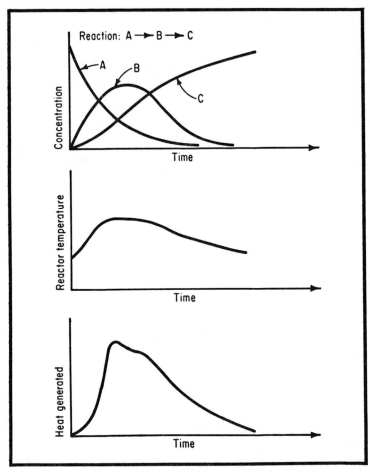

Fig. 1-24. Typical profiles of process variables in a consecutive reaction. Heat generated by the reaction is greatest during conversion of A to B, but the temperature of the reactor is controlled at an optimum setting to insure maximum conversion in a minimum amount of time.

operated at a constant temperature, using cooling coils inside the vessel, a cooling jacket, and/or external heat exchangers to remove the heat of reaction. An *adiabatic* reactor does not remove or add heat, relying instead on recycle streams or manipulation of reactants, catalysts and pressures to control the reaction.

Figure 1-24 illustrates typical time profiles of important process variables in a consecutive reaction process. In this example, ingredient A is charged into a reactor; after heating, it reacts to form product B which reacts to form product C. Figure 1-24 graphically shows the heat generated by the exothermic chemical reaction and the temperature of the reactor during the reaction process.

The reaction is controlled by manipulation of the reactor temperature. The temperature is initially increased to accelerate conversion of A to B; as the heat generated by the reaction increases, coolant is circulated to control the temperature at a particular level; after the concentration of C reaches a certain value, the reactor is allowed to cool. Any modification of the reactor temperature during the process would have an effect on the production of C. The effect could either result in a longer or shorter batch time, or a reduction in the amount of product C. During succeeding batch production runs, the optimum time and temperature sequences are duplicated, insuring maximum production of C.

BASIC CHEMICAL KINETICS

Chemical reactions are dependent on temperature. The variation in a specific reaction rate with temperature is usually described by the Arrhenius equation:

$$k = \alpha e^{(E/RT)} \tag{1.23}$$

where

k specific reaction rate (min^{-1})
α pre-exponential factor (min^{-1})
E activation energy of reaction (Btu/mole)
R perfect gas constant (1.99 Btu/mole-°R)
T absolute temperature (°R)

According to the Arrhenius equation, specific reaction rates are exponential in nature, increasing with temperature. The activation energy E of a reaction determines the degree of temperature dependence—the larger the value of E, the more temperature-dependent is the reaction.

Chemical reactions can follow quite complex paths and sequences, but for engineering purposes such as equipment design and control system analysis, most reactions can be considered as one of four types: irreversible, reversible, consecutive or simultaneous.

- *Irreversible* $A \xrightarrow{k} B$

Reactant A forms product B at a rate R_A that depends on the specific temperature-dependent reaction rate k and the concentration of A (C_A) at any point in time:

$$R_A = -k \, C_A \qquad (1.29)$$

The larger the value of k, the faster the reaction proceeds.

As reactant A is consumed, its concentration decreases and the reaction slows down. Therefore, towards the end of the batch cycle, less and less A is available to react. Essentially all of A can be converted into B if the batch time is long enough.

If two reactants (A and B) are involved, forming a product C:

$$A + B \xrightarrow{k} C$$

the reaction rate depends on the concentrations of both reactants:

$$R_A = -k \, C_A \, C_B \qquad (1.25)$$

- *Reversible* $\quad A \underset{k_2}{\overset{k_1}{\rightleftharpoons}} B$

Reversible reactions are those in which chemical component A can react to form product B, and B can also react back to form A. The forward reaction rate of A to B depends on k_1 and the concentration of A. The reverse reaction of B back to A depends on k_2 and the concentration of B. The net rate of consumption of A is:

$$R_A = -k_1 \, C_A + k_2 \, C_B \qquad (1.26)$$

Starting with pure A causes the forward reaction to proceed rapidly. But as the concentration of B builds up, the reverse reaction begins to occur. An equilibrium or steady-state condition will eventually occur where both A and B exist. The concentrations of A and B at equilibrium can be found by setting Equation 1.26 equal to zero, since at steady state there is no net consumption of A:

$$\frac{C_B}{C_A} = \frac{k_1}{k_2} = K \tag{1.27}$$

The equilibrium constant K indicates the maximum attainable conversion of A to B. A big K means a lot of product B will be formed at equilibrium; a small k means little B will be formed.

Thus, unlike irreversible reactions where A could be completely converted to B, reversible reactions have an inherent kinetic limitation which restricts conversion.

The specific reaction rates k_1 and k_2 both increase with temperature, which means that equilibrium conditions will be attained more quickly. However, the ratio of k_1 to k_2 decreases with temperature if k_2 is more temperature-dependent than k_1. In this case, the equilibrium constant K decreases with temperature, thus reducing the maximum attainable conversion. In this type of system there would be an optimum batch temperature profile for maximum production of B. Early in the batch a high temperature is used to get most of reactant A converted rapidly to B. As the concentration of B increases, the batch temperature is reduced to inhibit conversion of B back to A.

- *Consecutive* $A \xrightarrow{k_1} B \xrightarrow{k_2} C$

Two-step reactions are quite common in which A reacts to form B, which then reacts to form C. If B is the desired product, the reaction must be stopped when the concentration of B reaches its maximum (see Fig. 1-24). When the second reaction is more temperature-dependent than the first, there is, theoretically, an optimum temperature profile. High temperatures are used early in the reaction to rapidly convert A into

B. Lower temperatures are used later to prevent the loss of B

- *Simultaneous* $\begin{cases} A \xrightarrow{k_1} B \\ A \xrightarrow{k_2} C \end{cases}$

Another common reaction is one in which reactant A can form two different products, B and C. Reaction conditions must be adjusted to favor formation of the desired product. A high conversion of A does not necessarily mean a high yield of the desired product B. If k_2 is more temperature-dependent than k_1, an optimum temperature profile exists which starts at a low temperature, favoring the reaction to B, and finishes at a high temperature to complete the reactions.

Kinetic performance of a reaction is also determined by the type of reactor vessel. Three types of chemical reactors are commonly used: continuous stirred-tank reactors (CSTR), tubular reactors and batch reactors. Some commercial systems use multiple CSTR's in series or combinations of different reactor types. Recycle streams are also often used to improve overall conversion and/or to improve control, particularly in tubular reactors.

CONTINUOUS STIRRED-TANK REACTORS

A CSTR is a reactor into which feed is continuously introduced and from which product is continuously removed. Good mixing is usually achieved by mechanical agitation.

The steady-state performance of a CSTR can easily be calculated from a component balance on reactant A. As an example, consider an irreversible reaction of A to B. The feed rate to the reactor is F (ft³/min), and the concentration of reactant in the feed stream is C_A (moles/ft³). The volume of reaction mass in the reactor is V (ft³), and the specific rate constant is k (min⁻¹). Reactant enters in the feed at a rate F C_{AO}. It is lost in the product steam at a rate F C_A. It is consumed by the chemical reaction at a rate V k C_A. Since input minus output is equal to consumption:

$$F\ C_{AO} - F\ C_A = V\ k\ C_A \qquad (1.28)$$

Solving for C_A/C_{AO}:

$$\frac{C_A}{C_{AO}} = \frac{1}{1 + k \, V/F} = \frac{1}{1 + k \, \tau_p} \quad (1.29)$$

where τ_p is the holdup time in the reactor.

The smaller the C_A/C_{AO} ratio, the larger is the conversion of reactant A to product B. Equation (1.29) clearly shows that either k or τ_p must be large to achieve a high conversion in a single CSTR. A slow reaction having a small k will require a very large volume to achieve reasonable conversions.

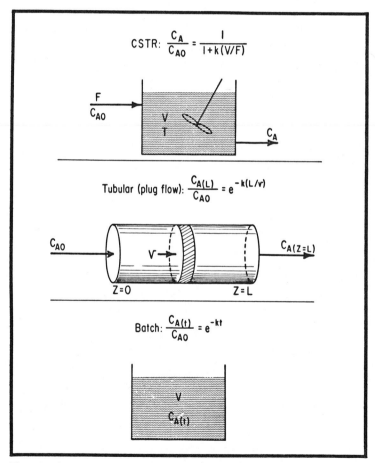

Fig. 1-25. Conversion performance equations of CSTR, tubular and batch reactors for the same reaction.

TUBULAR REACTORS

A tubular reactor is also a continuously operated reactor, but instead of trying to achieve good mixing, the opposite is desired. Tubular reactors typically are long thin tubes: feed comes in one end, and the reaction proceeds as the material flows down the length of the tube, more or less in plug flow. Temperature and compositions vary with both time and axial distance (z) down the reactor. Radial variations in temperature and compositions inside the tubes can also exist if large amounts of heat are liberated by the reaction and the tubes are large in diameter.

Assuming ideal plug flow of material down the tube at a velocity v (ft/min) and constant temperature, a steady-state component balance on a small slice of reactor length dz, Fig. 1-25, leads to an ordinary differential equation:

$$\frac{d\,C_A}{dz} = -\,k\,C_A/v \tag{1.30}$$

Integrating from the inlet ($z = 0$) to the exit ($z = L$) of the reactor gives:

$$\frac{C_{A(L)}}{C_{AO}} = e^{(-L/v)} = e\,(-k\tau_p) \tag{1.31}$$

where τ_p is the residence time in the reactor.

If k is small, a long tubular reactor is required. Equation (1.31) indicates that plug velocity in the tubular reactor could be slowed to compensate for a small k, but this results in a large number of tubes, poor flow distribution among the tubes, and poor heat transfer. Also, at low velocities flow becomes laminar and the plug-flow assumption no longer applies.

The most serious problem in tubular reactors is controlling temperature down the length of the reactor. Severe local hot spots can occur and temperature control is often so difficult that an inert recycle stream must be used to provide a heat sink. The reactor is broken up into sections that operate adiabatically, and enough recycle is used to prevent the adiabatic temperature rise from becoming too large. Interstage cooling in heat exchangers or adding more cold recycle is used to control each section at some optimum inlet temperature.

BATCH REACTORS

In a batch reactor, the vessel is initially charged with a volume of reactant; the volume remains constant through the reaction process, and no inflow or outflow terms enter into the dynamic equation. The component balance equation, therefore, is an ordinary differential equation with time as the independent variable:

$$dC_A/dt = -kC_A \qquad (1.32)$$

If the initial concentration of reactant at time equals zero is C_{AO}, and k is assumed to be a constant (implying an isothermal process), integration gives:

$$\frac{C_{A(t)}}{CAO} = e^{(-kt)} \qquad (1.33)$$

Notice the similarity between Equations 1.31 and 1.33. A plug-flow tubular reactor is described by the same equation as an isothermal batch reactor. This means that if the residence time τ_p in an isothermal tubular reactor is the same as the batch time of an isothermal batch reactor, the kinetic performance of the two reactors will be identical.

The basic kinetic performances of the three types of reactors are given by Equations 1.29, 1.31 and 1.33. Figure 1-26 is a plot of the C_A/C_{AO} ratios for various values of $k\tau$ in a simple irreversible reaction for a single CSTR and a batch reactor. The tubular reactor curve is the same as for the batch reactor. This plot clearly shows the inherent kinetic advantage of a batch reactor over a CSTR. Multiple CSTR's in series improve performance, but several stages must be used to approximate the performance of a batch reactor.

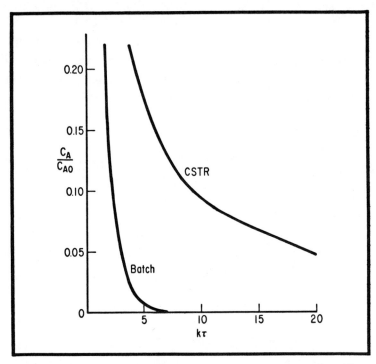

Fig. 1-26. A comparison of conversion performances, C_A/C_{AO}, at various of $k\tau$, shows the inherent efficiency advantage of a batch reactor over a CSTR. For a given reaction and reactor volume, the batch reactor has a smaller reaction time, and can produce the same amount of product faster than a CSTR.

The same comparison holds for most types of reactions, with the exception of autocatalytic reactions where higher product concentrations give higher reaction rates.

Batch reactors tend to be used when k values are very small, when the process involves many reaction steps, when there are difficult material-handling problems with continuous feeds, when isolation and sterility problems occur, when molecular weight distributions in polymerization reactions are adversely affected by the residence time distribution of a CSTR, or when production rates are not very large.

DYNAMIC STABILITY OF BATCH REACTORS

The basic feedback control loop in a batch reactor is designed to manipulate heat-removal equipment for control of reactor temperatures. The process open-loop transfer function, relating temperature to heat-removal rate often contains a positive pole—a point where a root of the denominator is greater than zero—which indicates open-loop instability in the reaction:

$$\frac{T(s)}{Q(s)} = G(s) = \frac{K_p}{\tau_p s - 1} \qquad (1.34)$$

where

T = temperature
Q = heat-removal rate
K_p = process steady-state gain
τ_p = reactor time constant
s = LaPlace transform variable

Open-loop instability can occur when the reaction is exothermic and irreversible and the reaction heat generation exceeds the capacity of the heat-removal equipment. The larger the heat of reaction or activation energy and the smaller the heat transfer area or coefficient, the more possible is open-loop instability.

An open-loop unstable process can be stabilized by adding a feedback controller which will increase the heat-removal rate Q as temperature increases. The closed-loop characteristic equation (CLCE), with the proportional-only controller function $(B(s) = K_c)$ added, is:

$$1 + B(s)G(s) = 1 + K_c \frac{K_p}{\tau_p s - 1} = 0 \qquad (1.35)$$

A root locus plot of this equation is shown in Fig. 1-27A. A root locus plot shows the real and imaginary roots of a closed-loop characteristic equation as a function of feedback controller gain. The roots are plotted on real and imaginary axes in the s-plane.

As controller gain K_c increases from zero, the single root of Equation 1.35 moves left along the real axis. When the value of K_c reaches the value marked $(K_c)min$, the root enters the stable left-hand side of the s-plane and the system becomes closed-loop stable. For any values larger than $(K_c)min$, the system is stable; for any values lower than $(K_c)min$, the system is closed-loop unstable.

Equation 1.35 describes a simple, ideal system where both temperature measurement and heat removal are instantaneous. In a real reactor, however, temperature measurement lags can be significant—especially in polymerization reactors where thermowells can become coated. The dynamics of the cooling system (coil, jacket, heat exchangers, etc.) and their inherent response delays also cannot be neglected. The process transfer function of a real reactor has several lags, or holdup times, in addition to the single positive pole due to the reactor time constant.

The closed-loop characteristic equation of a real reactor with a time constant (τ_{p1}), and delay lags in measurement (τ_{p2}) and heat removal (τ_{ps}) is:

$$1 + B(s)G(s) = 1 + \frac{K_c K_p}{(\tau_{p1}s - 1)\,(\tau_{p2}s + 1)\,(\tau_{p3} + 1)} \quad (1.36)$$

The root locus plot of this equation, Fig. 1-27B, shows that the system is "conditionally stable;" i.e., there is a range of controller gains between $(K_c)min$ and $(K_c)max$ for which the system is closed-loop stable. If the negative poles, $-1/\tau_{p2}$ or $-1/\tau_{p3}$, move to the right along the real axis due to increases in lag times, the root locus plot also moves to the right and the range of $(K_c)max$ becomes narrow. The reactor could become uncontrollable.

Another common control problem in batch reactors is reset windup. If integral action is used in the controller, its output signal will saturate, if left on "automatic," during those periods when the reactor is being charged or unloaded and the temperature controller is not in use. A temperature overshoot will occur when the process comes back on control since the

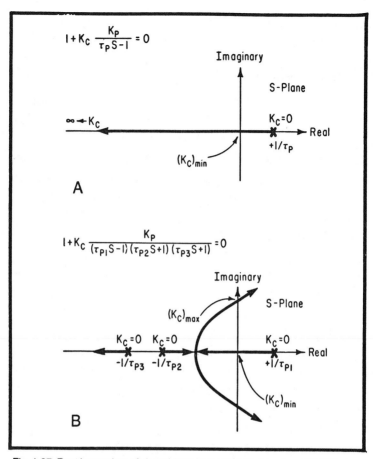

Fig. 1-27. Root locus plots of closed-loop characteristic equations (CLCE) show the roots of the equation plotted along real and imaginary axes on the s-plane. Root values to the right of the imaginary axis indicate closed-loop instability; values to the left indicate closed-loop stability. A) The locus of the single root of this equation lies entirely along the real axis. For any value of K_c to the left of $(K_c)_{min}$, the loop is stable. B) Root values for a system with three delays included in the CLCE include three poles. In the normal operating range, the controller gain lies between $(K_c)_{min}$ and $(K_c)_{max}$ and the loop is stable; outside these limits, the plot enters the right-hand side of the s-plane and the loop becomes unstable.

error must change sign before the integral term can start to decrease. Reset windup can be prevented in analog equipment by using special "batch" controllers, which stop the integral action when windup is detected, or by using "external reset feedback" (Ref. 1.25). It should be noted that simply limiting the controller output reduces reset windup but does not eliminate it. Digital computer control algorithms should also be able to handle this windup problem.

OPTIMAL CONTROL

In theory, many batch reactors have an optimum time-temperature profile that will maximize product yield in a minimum amount of batch time. Theoretical studies have appeared in the literature for many years discussing this problem, applying such sophisticated mathematics as the calculus of variations, the "Maximum Principle," dynamic programming, etc.

As pointed out by Marroquin (Ref. 1.26), there is considerable incentive to apply optimal control when exothermic reversible reactions are involved. A temperature profile that decreases with time can save significant amounts of batch time (25-50 percent and more), particularly when high conversions are desired.

On the other hand, Marroquin also showed that optimal control yields very little reduction in batch times over a constant temperature (best isotherm) for consecutive, irreversible reactions. Yield improvements of no more than three percent were typical for a wide range of kinetic parameters. These results apply to two consecutive reactions such as A to B to C; a batch cycle with more steps may well have different "best isotherms" for the various phases of the operation.

The degree of control of either an optimal or a best isotherm profile is dependent on the ability of the heat removal equipment to control temperature effectively. The amount of control and speed of response available is frequently a function of the complexity and cost of a cooling system. Figure 1-28 shows a variety of common batch reactor heat removal systems. In some cases, especially if reactor temperatures are significantly above 100°C, the coolant must be something other than water and a closed, recirculating system is required.

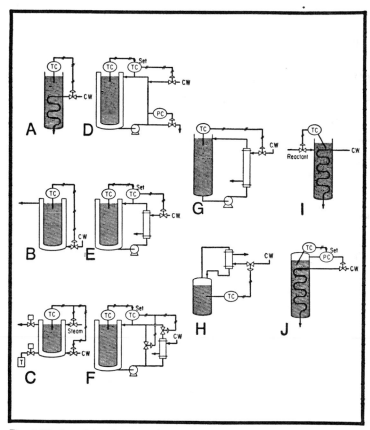

Fig. 1-28. A variety of commonly used batch reactor heat removal control systems. A) Cooling water flows through internal coil in reactor. B) Coolant flows through external reactor jacket. C) Split-range system, where steam heats reactor up to desired temperature and cooling water then removes heat of reaction. D) Cascade temperature control of cooling water additions to a recirculating coolant loop reduces cooling jacket dynamic lags. E) Cascade loop similar to D, but a heat exchanger cools the recirculated coolant. An additional lag is introduced by the heat exchanger. F) A two-valve bypass system provides very rapid control of jacket temperature. This is a cascade system, combining elements of D and E. G) Instead of circulating coolant through a jacket, the reactant liquid itself may be pumped through a heat exchanger. Large amounts of heat transfer area are realized, but circulation rate must be high to achieve good control and mixing. H) An autorefrigerated or boiling-liquid reactor, where heat of reactor is removed by vaporization of liquid in the reactor, has extremely fast heat transfer dynamics. Vapor is condensed and cooled liquid returned to the reactor, normally by gravity flow through a U-seal. I) Batch reactors are sometimes run in a semicontinuous manner. Reactant is fed into the reactor as rapidly as possible by the temperature controller while maximum cooling is maintained. In some systems, this is the best solution to a limited heat-removal capacity. J) A cascade temperature-to-pressure system controls both temperature and pressure. This system can be used in a closed reactor if pressure changes can be detected more rapidly than temperature changes.

Some heat-removal systems pump the reactant itself through a heat exchanger, Fig. 1-28G, or remove heat by vaporizing reactant liquid, Fig. 1-28H. In a closed system, the temperature can be controlled in cascade by manipulation of coolant flow based on pressure measurement (Ref. 1.27) as shown in Fig. 1-28).

Each heat-removal system has its own penalties, either in terms of time lags for the cooling process or in increased cost due to the system's complexity.

REFERENCES

1.25 Buckley, P.S., "Designing Override and Feed-forward Controls," *Control Engineering*, Part I, August 1971, pp. 48-51; Part II, October 1971, pp. 82-85.

1.26 Marroquin, G. and Luyben, W.L., "Practical Control Studies of Batch Reactors Using Realistic Mathematical Models," *Chemical Engineering Science*, Vol. 28, 1973, pp. 993-1003.

1.27 Hopkins, B. and Alford, G. H., "Temperature Control of Polymerization Reactors," *Instrumentation Technology*, May 1973, pp. 39-43.

BIBLIOGRAPHY

Levenspiel, O., *Chemical Reaction Engineering*, John Wiley & Sons, 1962.

Mayer, F.X. and Spencer, E.H., "Computer Simulation of Reactor Control," *ISA Journal*, July 1961, pp. 58-64.

Millman, M.C. and Katz, S., "Linear Temperature Control in Batch Reactors," *Industrial & Engineering Chemistry Process Design and Development*, Vol. 6, October 1967, pp. 447-451.

Schrock; L.J., "Minimum-time Batch Processing," *ISA Journal*, October 1965, pp. 75-82.

Shinskey, F. G. and Weinstein, J.L., "Dual-Mode Control System for a Batch Exothermic Reactor," Paper No. 6-4 1-65 *Advances in Instrumentation*, Vol. 20, 1965, Proceedings of the 20th Annual ISA Conference and Exhibit, Los Angeles

Siebenthal, C.D. and Aris, R., "Application of Pontryagin's Method to the Control of Batch and Tubular Reactors," *Chemical Engineering Science*, Vol. 19, 1964, pp. 747-761.

WILLIAM L. LUYBEN is a Professor of Chemical Engineering at Lehigh University, Bethlehem, PA. Article is based on a paper presented at the ISA Joint Spring Conference, Montreal, 1975.

Batch Control Problems

R.E. BEST

PART I: COPING WITH CONTROLLABILITY

Bringing a batch to the desired endpoint state within the constraints of dynamics and available hardware can force nonstandard measurement and control techniques upon the instrumentation engineer. In common applications, the control system receives a set of measurements (e.g., temperatures, pressures, flow rates, time), processes them according to previously programmed instructions and makes calculations based on a set of control variables with empirically derived procedures which constitute, for example, a PID algorithm.

For the most part, engineers have little doubt that the computer (or equivalent control system) can generate valid output data from a given set of input data. However, the controllability of a process depends not only on the type of control system designed for it, but also on certain characteristics of the process itself—particularly its dynamic response to disturbances and manipulated inputs. Controllability is also a function of the set of process variables selected for measurement and manipulation by the control algorithm.

MEASUREMENT DIFFICULTIES

In batch processes it is customary to frequently fill and empty tanks, reaction vessels, weighing vessels and pipes, to frequently start and stop conveyor belts and to execute other operations intermittently as well. If the products are viscous, sticky or tend to crystallize, it is highly likely that pipes will clog, valves will stick, pH electrodes will become encrusted and so on. When the next batch operation starts with fouled equipment, the control system may receive incorrect feedback as to the state or condition of the processing equipment. This misinformation can lead to incorrect sequencing action. Therefore, in batch systems, it is a recommended practice to provide feedback confirming the execution of every drive command.

Another characteristic of batch operations is that many of the process variables are difficult—or even impossible—to measure. The following problems are often encountered:

- The sensor signal does not truly represent the desired measurement. For example, the voltage delivered by a redox electrode not only depends on the particular reduction/oxidation reaction being measured, but also on the sum of all electrochemical potentials of all redox reactions occurring in the measured system.

- Some sensors are rendered unreliable while others can be affected by systematic errors due to such conditions as dirt and crust formation. Sensors like pH electrodes, capacitive travel transducers and ultrasonic level switches are particularly sensitive to such fouling effects.

- The following types of interference can be superimposed on electrical measurement signals: hum, amplifier noise and cross-talk, all of which are due to stray inductive and capacitive coupling or to improper grounding and shielding of cables.

- Measurements do not have the required degree of accuracy for the intended control purpose.

Moreover, the controllability of a batch chemical plant is often poor because important information is missing entirely. There are no general cures for these measurement problems. The methods which follow however, have proven useful in practice.

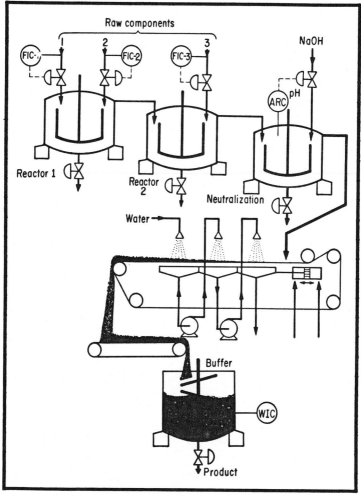

Fig. 1-29. Mass balance calculations, based on the total amount of the three feed components in this dyestuff process, facilitate measurements of water in the product.

The null method—In many cases where a physical variable is difficult to measure, it may suffice simply not to measure it, especially if this variable is irrelevant to the control strategy. Potential users of process control computers are often overwhelmed by their seemingly inherent infinite capabilities and therefore tend to measure everything in sight, including the copper losses in the main transformer!

Cross-correlation method—A physical variable that is difficult to measure can often be inferred or calculated from simpler, alternative measurements.

We employed this method in a dyestuff production plant, Fig. 1-29, where measuring the humidity of the dyestuff paste after filtering was a problem. Dyestuff is synthesized by proportional blending of three dry feedstreams whose flow rates are measured, followed by spray water addition. To determine the water content of the paste product, we used an existing plant computer for mass balance calculations. Water content is calculated by subtracting the total amount of the three feed components in the mix from the weight of the dyestuff in the buffer vessel.

Another problem of nonspecific measurements arose in connection with a batch endpoint determination. Many chemical reactions (for instance, reductions, oxidations and diazotizations) are characterized by an electrochemical or redox potential. In batch operations involving such reactions, the redox potential reaches a minimum or maximum when the reaction is completed. Since the measured potential is the sum of all electro-chemical potentials existing in the reaction, the measured value is shifted accordingly by any impurity present, and its final value cannot be predicted.

In addition, the setpoint cannot be initially placed at an assumed maximum or minimum value because the chemical equilibrium of the reaction would be affected, resulting in an undesirable ratio of product and by-product. Therefore, the setpoint must be raised incrementally until it reaches a maximum, Fig. 1-30.

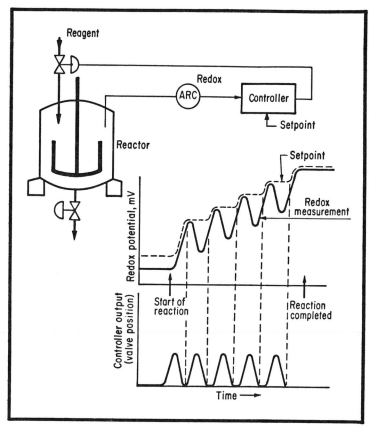

Fig. 1-30. Since the actual maximum value cannot be determined before the reaction starts, the setpoint is raised in increments by tracking the measured value with a peak follower.

This control action can be accomplished by either hardware or software. The control scheme shown in Fig. 1-30 was implemented by a hardware controller developed by the author. A built-in peak follower circuit tracks the redox potential measurement, so that the controller setpoint is not a constant value, but reflects the maximum value of the measurement. In this manner, the controller setpoint is changed so that the desired reaction can occur with a minimum number of side reactions. With this controller, some dyestuff reactions can be performed in only 30 percent of the time previously required.

OVERCOMING UNRELIABLE MEASUREMENTS

In those cases where a measurement device cannot provide the required degree of accuracy or reliability, the problem can sometimes be solved by measuring another physical variable which suitably describes the system. Examples of possible substitutions are:

- For flow rate—weight, volume or Doppler effect measurements.
- For humidity—mass balance calculations as in the example above.
- For liquid level—weight measurements; or empirical (uncalibrated) measurements of static pressure, capacitance or attenuation of nuclear radiation.

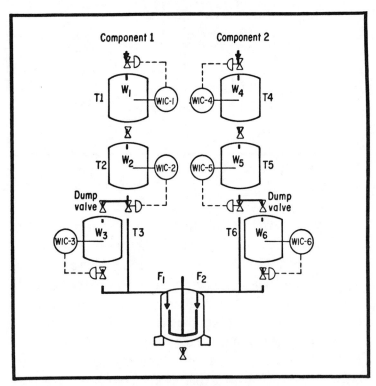

Fig. 1-31. Conventional control with flowmeters in this dyestuff process would not provide the required accuracy in ratioing components 1 and 2. An alternative method involves weighing the contents of vessels 1, 2, 4 and 5 with load cells.

- For density (as an index of quality)—boiling temperature, velocity of sound, resonant frequency (tuning fork method).

Substitution of weighing for flowmetering is illustrated by a computerized, semicontinuous, dyestuff production line that synthesizes product by charging two liquids stoichiometrically into a cascade reactor. A ratio control scheme based on currently available flowmeters would not provide the required accuracy over extended periods. So an arrangement which weighs the contents of vessels mounted on high-precision electronic load cells was designed, Fig. 1-31.

The timing waveforms shown in Fig. 1-32 demonstrate that steady component streams F_1 and F_2 can be produced by repetitively filling and emptying the series-connected vessels. The emptying process is accomplished by downramping the setpoint of controllers WIC-2 and WIC-5, respectively. In addition, the amounts of each ingredient in every batch are measured twice: first in weighing vessels T_1 and T_4, and then in vessels T_2 and T_5. Thus, the proper functioning of the load cells and the weighing electronics is checked automatically.

Weighing vessel T2 cannot accept the contents of T1 and simultaneously check this weight while dispensing liquid to the synthesis reactor. Therefore, auxiliary vessel T3 is added to the train. Toward the end of a discharge ramp, at a preset weight value, the contents of T2 are quickly discharged or dumped into T3. The auxiliary vessels T3 and T6 thus feed product to the synthesis reactor during the last part of the discharge cycle. These "dribble" flows are also controlled by downramping the weight setpoint.

Duplication of sensors (load cells in this case) is a commonly used method to increase the reliability of a system. A dual sensor system is easily capable of detecting when one of the two sensors has failed, but cannot indicate which one it is. When this additional refinement is needed, a triple sensor system must be used.

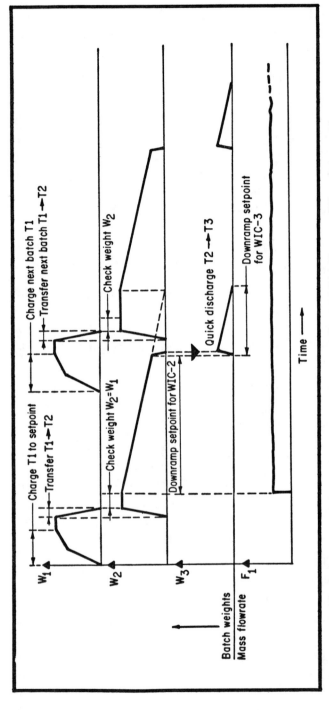

Fig. 1-32. This timing diagram illustrates the operation of the system shown in Fig. 1-31. Weighing vessel T2 cannot accept the contents of T1 and simultaneously check this weight while dispensing liquid to the synthesis reactor. Therefore, auxiliary vessel T3 is added to the train. Toward the end of a discharge ramp, at a preset weight value, the remaining contents of T2 are quickly discharged or dumped into T3. The auxiliary vessels T3 and T6 thus feed product to the synthesis reactor during the last part of the discharge cycle. These "dribble" flows are also controlled by downramping the weight setpoint.

94

COMPENSATING FOR PROCESS DYNAMICS

Even with reliable hardware and software configurations, the control of a process may be poor because its dynamic behavior is difficult to handle. According to state space theory, the status of a control system can be described by a set of physical quantities represented by an n-dimensional plant state vector $x(t)$. When a batch operation is performed, the system is manipulated by the application of a set of control variables forming the so called control vector $m(t)$, in such a manner that the plant state vector is moved from an initial point x_o to an endpoint x_e. The transition from x_o to x_e can take place on an infinite number of paths. But there exists an optimum trajectory, over which the system is optimized with respect to a particular performance function—for example, minimum energy or minimum transition time.

In real life, the control engineer makes only very limited use of such advanced theory. When working on a plant project, he is satisfied to measure the state of the plant with reasonable accuracy and to know—at least empirically—how the system responds to the control variables. Knowledge of static behavior is almost trivial in many cases; for example, in a temperature control loop the temperature will ultimately rise when heat is applied to the system. However, process dynamics can severely impair the controllability of the system, especially if the loop contains large deadtimes.

An example of difficult dynamics in a dyestuff process involves accurate control of product dryness in a spray drier, Fig. 1-33A. The outlet air temperature is normally taken as a measure of product dryness. Any deviation from the desired outlet temperature alters the rate of the feed pump. However, due to the long deadtime from air inlet to outlet, any disturbance in the inlet temperature is detected too late. This control system is therefore prone to instability and overshoot; and start-up of the process is particularly cumbersome.

An alternative approach, based on mathematical models developed by Sandoz makes use of feedforward techniques,

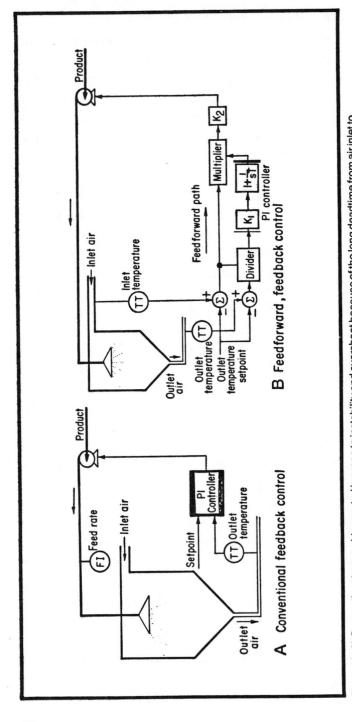

Fig. 1-33. A) Conventional spray drier control is prone to instability and overshoot because of the long deadtime from air inlet to outlet, where exit air temperature is taken as a measure of product dryness. B) This feedforward control scheme compensates for inlet temperature disturbances by making pump speed adjustments in addition to those based on outlet temperature feedback.

A Conventional feedback control

B Feedforward, feedback control

Fig. 1-33B. In this configuration, the difference ΔT between measured inlet and temperature and desired outlet temperature is used to precalculate an approximate feed rate $F_o = K_2\Delta T$. Any disturbance in the inlet temperature is instantaneously compensated by this feed-forward control loop. To insure that the outlet temperature will reach its correct value, the conventional PI feedback loop is retained, and adjusts the pump speed based upon outlet temperature deviations from setpoint. But its gain must be adjusted for compatibility with the gain K_2 in the feedforward path.

With the original configuration, the drier could be started up only by very careful upramping of the inlet temperature, taking approximately one hour. The modified system can be started up with full heat within two minutes.

PART II: OPTIMIZING SAFETY AND RELIABILITY

Safety considerations should be an integral part of the process design and equipment selection—even before basic control strategies are planned. Attempting to achieve the desired level of safety by assigning this responsibility to a particular control scheme is like putting the cart before the horse.

A system is considered safe when personnel, machinery and material assets are protected from damage in the event of failure or malfunction of process equipment, instrumentation or plant services. A process engineer should start the layout of a chemical product line with a careful analysis of the areas that may be potentially hazardous. Dangerous conditions can arise at various points in a batch process when:
- The chemical reaction is exothermic
- A critical component of the mechanical equipment fails (e.g., a stirrer breaks or a cooling pump fails)
- The sensing instrumentation fails
- The control system breaks down (e.g., a controller malfunctions or a valve sticks)

- Power or other utility (e.g., electricity, water, air pressure) is lost.

SAFETY BY PLANT DESIGN

Once these potentially hazardous conditions have been determined, they should be precluded by either preventive or event-oriented measures in the design. In every process design, it is preferable to include measurements that will *predict* unsafe conditions so that preventive actions can be taken, as opposed to those that sense the *occurrence* of undesirable events, such as the overheating of a reaction mixture. (Prohibiting smoking is undoubtedly better than extinguishing a fire!)

If the safety of a production line depends on continuous measurements, the sensors performing this critical function should have backup units. In processes where safe operation is possible only when a mixer or other electrically-powered equipment is running, autonomous electric power generators should be provided to maintain process stability. Alternatively, coolant reservoirs can be provided to quickly bring a reaction mass to a safe temperature if the power fails. Safety considerations should be reflected in all aspects of batch process design, including such in-line components as valves, instruments and the process control computer.

The choice of process valving to ensure safety depends on the characteristics of the batch operation under control. Most types of valves on process equipment have a preselected emergency safe position or state. For example, product valves on a reaction vessel are generally closed in the failsafe condition, whereas cooling water or quenching valves are designed to fail open. The point is that the failsafe position of each valve in the process must be predetermined by carefully analyzing process operations and interactions.

Fortunately, even though they may not be chosen for safety reasons, many actuators and valves go into a predetermined safe state when control or drive power is lost.

COMPUTER-ADDED SAFEGUARDS

Under the control of a reliable process control computer and program, a chemical batch process can be operated with an enhanced degree of safety. Production line units, such as a reaction vessel, size reduction mill or drier, are often controlled by a set of sequencing programs that run in parallel, Fig. 1-34. Normal batch processing is controlled by one or several "normal operating" sequences. In parallel to these programs, one or more safeguarding sequences check critical plant equipment for alarm conditions. When a safeguard sequence program detects an alarm condition, it immediately triggers an emergency sequence program which puts the process in a safe state and alerts the operator.

What happens, however, if the computer fails? Many solutions have been proposed. One of the simplest, most efficient methods is a failure-detecting program that executes once every second. The program continuously checks the correct operation of arithmetic registers, A/D and D/A converters, I/O channel addressing, and so on. As soon as a

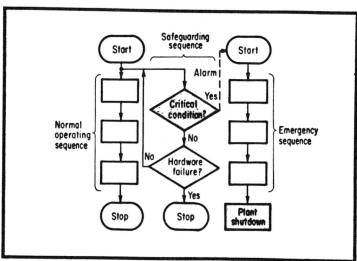

Fig. 1-34. Safeguard software routines, executed concurrently with the normal operating sequences, check plant equipment for alarm conditions. An immediate branch is made to an emergency sequence when such a situation occurs. If a critical hardware failure occurs, the safety program shuts the plant down.

failure is detected, a hardware alarm signal is generated; a relay is then de-energized, and the resulting signal disconnects the process from the computer interface and drives the plant to a safe state. Of course, this necessitates some additional logic circuitry between computer interface and process.

At a new Sandoz Ltd. plant, this type of safety logic circuitry has been incorporated in the digital output interface of the computer, Fig. 1-35. During normal computer operation, the computer failure relay is held closed by the failure detection program. Thus, the C-bus is enabled with a logic 1 and the digital output of the computer directly controls the process. However, when the computer fails, this relay drops out, the S-bus is enabled and the process is then controlled by a bank of preset safety switches.

If the operator wishes to bring the process under manual control, he throws the manual backup switch into the MANUAL position, so that he can drive the actuators with the remaining bank of switches, marked "Manual operation." The outputs of the logic gates feed solid-state relays that are capable of driving valve actuators or other AC loads directly. This arrangement avoids troubles due to relay contact burnout or oxidation.

In direct digital control (DDC), the safety of each control loop when the computer fails must be considered individually. Several different schemes have been used by our company to make DDC loops failsafe. In some cases, it suffices simply to disconnect the appropriate control loop from the computer with a relay contact. Under other circumstances, the computer is set up in normal plant operation to perform an incremental control algorithm and to feed pulses to an electromechanical I/P converter containing a stepping motor. In the event of computer failure, further pulses are cut off and thus the pneumatic output is frozen at its last value. This action affords enough safety in many applications. However, in critical situations, where maintaining continuous control of the loop is a

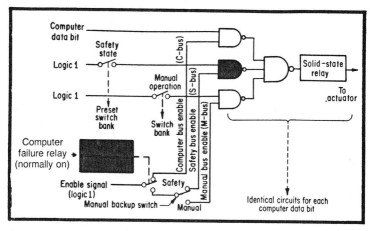

Fig. 1-35. If the computer fails, the failure relay is de-energized and enables control of the batch by the bank of preset safety switches. The operator can take over direct manual control by setting the back-up switch to MANUAL. Note that the circuitry shown here represents that portion associated with only one data bit.

strict requirement, autonomous analog backup controllers are used.

Even when the computer operates correctly, a dangerous situation can occur when, for example, a temperature transducer or transmitter fails. To overcome this problem, autonomous safety interlocks are often inserted at critical places in the system. These interlocks are hardware devices, often in the form of "burnout relays" or similar auxiliaries connected to analog transmitting or recording instruments. Hence, the control system eventually becomes a hierarchy, Fig. 1-36. Safety interlocks are assigned first priority since they can override the other two levels shown.

Sandoz does not believe that redundant computers materially increase the safety of a system, since they have similar parts that could fail simultaneously or in quick succession as, for example, if an overload is switched from one to the other. They also usually have some parts in common, whose failure would shut down both computers.

COUPLING RELIABILITY WITH SAFETY

The control engineer generally must ensure that a plant is

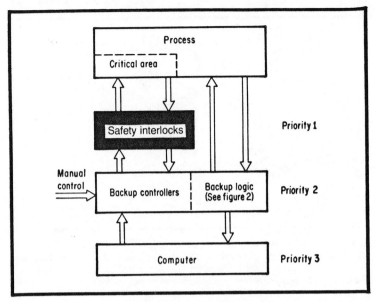

Fig. 1-36. Backup control for safety and reliability is established in a hierarchy of priority levels, with autonomous safety interlocks taking the highest priority. Other forms of backup may be analog, digital, or manual.

highly reliable as well as safe. System reliability may be increased or decreased by the procedures chosen to enhance system safety, depending on the process. In an extreme example, quenching a reactor at the slightest sign of trouble—such as sensor failure or loss of power—would definitely improve safety, but would so deteriorate reliability that the plant might well become uneconomical to operate. The control engineer is often confronted with the uncomfortable situation of trading safety for production.

In chemical product lines, component failures occur quite often, but do not necessarily result in the shutdown of the entire plant. Reliability can be defined as the ability of the system to withstand minor component or subsystem breakdowns. A quantitative measure of reliability can be expressed as availability:

$$\text{Availability} = \frac{\text{MTBF}}{\text{MTBF} + \text{MTTR}} \times 100 \text{ percent}$$

where

MTBF = mean time between system failure

MTTR = mean time to repair

Typically, availability must be more than 90 percent for the plant, and more than 99 percent for the control system.

In a computerized chemical plant, the computer is by far the most reliable subsystem, followed in decreasing order of reliability by:

1. Transmitters (electronic, pneumatic)
2. Sensors
3. Control valves and actuators
4. Plant equipment (mills, pumps, filters).

Further increasing the reliability of the computer system would not contribute much to overall reliability unless better plant equipment is installed. Dual computer systems are justifiable, we feel, only if extensive use is made of their combined higher capabilities, such as for background processing, on-line program assembling, debugging and editing, data logging and other functions.

Extensive studies have shown us that overall performance of a chemical plant can only be improved by redundancy in the plant equipment. A statistical computer model of redundant production lines based on Monte Carlo methods has been developed at Sandoz. Given values of failure probabilities for the individual plant components and given the degree of redundancy for the various components and their mean time to repair, the computer program calculates overall system availability. The model also helps to predict the holding capacity or tankage for intermediate product required to maintain uninterrupted operation.

FUTURE TRENDS

The complexity of chemical batch operations will increase considerably in the near future. Increasing demands for higher product quality, pollution abatement, and reduced energy consumption all necessitate tighter control, hence better controllability of the process.

Several studies of these objectives, both in the hardware and the software area, are continuing at Sandoz. One group is concerned with developments in the field of on-stream analysis. Much still has to be done in obtaining information about the chemical state of a reaction. Most of the problems involve getting representative samples with probe pickups and the associated sample conditioning required by an analytical instrument.

Another study deals with remote multiplexing of plant data. In a conventional layout, thousands of cables crisscross a plant. Aside from high cable costs, the engineering time spent in determining optimal cable routes is tremendous. Hardware engineering and installation could be greatly simplified by routing a single data bus or data highway around the plant, Fig. 1-37, and connecting every I/O element to this bus via the shortest route.

Every I/O device in this system, however, has to be equipped with associated logic circuitry for address decoding the data transfer control (InTech, January 1971, p. 63). The signal conditioner (transmitter and amplifier), the A/D converter and the power supply for each analog input must be located directly at the point of measurement. This equipment must be explosionproof in many cases because of its close proximity to the process. However, in view of the developments in electronics, this should not create major difficulties for the designer.

The appearance of the microcomputer on the market will certainly have an impact on automated plant operations. We think subdividing a production line into a number of independent plant areas, with each area having its own microprocessor and its own data bus, is feasible. In this configuration, the microprocessors may be able to perform DDC and sequencing jobs autonomoulsy; however, they have limited storage capa-

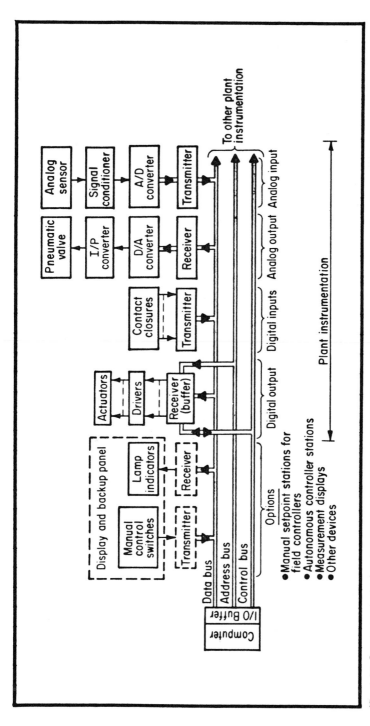

Fig. 1-37. Routing a single data bus around the plant greatly simplifies engineering design and installation efforts. Each I/O element in the system is connected to the bus through its own addressing and conversion logic.

105

city and would have to receive their sequencing programs in blocks from a master computer. At least one Swiss electronics company is working on such a system.

In the software area there is room for much improvement. Several manufacturers are now offering process control computers that have a suitable interface to the process, man-machine communications, and their own problem-oriented, high-level languages. But the user is sometimes forced to make modifications—with the aid of the manufacturer—to the system software. From our experience, the learning period required to tackle such changes is two years and must be repeated whenever a new computer system is installed.

Therefore, we are eagerly awaiting an "omnipotent" process language which is independent of hardware configuration and is easily adapted to commonly used computers. We are currently evaluating a process language PEARL (process and experiment real-time language) developed by European universities in conjunction with Swiss and German computer makers.

ROLAND E. BEST is Manager of the Control Engineering Dept. of Sandoz Ltd., Basel, Switzerland. Article is based on a paper presented at the ISA Joint Spring Conference, Montreal, 1975.

Section II—Measurement

Measurement techniques and devices constitute a major area of instrumentation and control engineering. This section of the book devotes eight chapters to this important subject. Theoretical aspects of measurement as well as specific techniques are dealt with.

A two-part chapter on humidity measurement surveys "commonly used (sensors) in industrial automated systems" and "only those sensors in wide use and for which error analysis have been published by the National Bureau of Standards." Part I describes humidity and the basic parameters by which it is measured. Part II then compares acutal humidity sensors.

Temperature measurement techniques also are reviewed. Part I focuses on compensation methods that minimize the effects of parasitics and offsets on common industrial sensors and multiplexers. Part II then details shielding methods to reduce common-mode errors in temperature measurements and looks at digital techniques to linerize temperature indications.

Information on selecting flowmeters, measuring flow with radiotracers, measuring force with miniature capacitive transducers, measuring torsional vibration, and understanding errors in automatic weight filling also is presented.

This measurement information provides a general coverage of some of the latest techniques and devices.

Review of Temperature Measurement Techniques

D.M. MACKENZIE and W.E. KEHRET

PART I: OPTIMIZING THE SENSOR/INSTRUMENT INTERFACE

Recent developments in integrated circuit (IC) technology now make it possible to overcome some of the traditional problems associated with conventional temperature sensors. Although few new temperature measuring elements have emerged in the last decade, such IC chips as operational amplifiers and multiplexers are available to 1) compensate for sensor nonlinearity, 2) reduce offsets, 3) handle low-level sensor signals and 4) process numerous inputs from various types of sensors with minimal hardware.

The wide variety of temperature sensors currently on the market suggests that there is no ideal device which can provide a linear, high-level output over an extremely wide temperature range in the industrial environment. It is often difficult to choose the appropriate sensor for a particular application because there are many criteria on which to base the decision:

- useful temperature range
- sensitivity
- cost and availability

Table 2-1. Temperature Sensor Trends and Operating Ranges

Sensor classification	Percent of market 1973	1980	Operating range, °C
Active sensors (no excitation)			
Thermocouples	56	50	−273 to +3,000
Radiation pyrometers	<5	<5	+1,000 to +4,000
Passive sensors (excitation required)			
RTDs	16	27	−273 to +1,000
Thermistors	28	23	−273 to +250
Semiconductor junctions	<5	<5	−273 to +150

- resistance to corrosive environments
- stability and accuracy
- response time
- complexity of measurement instrumentation required by the application
- maturity of technology.

At the present time, the thermocouple is the most common device for temperature measurement. Resistance temperature detectors (RTDs) have been popular in Europe for some time, and are rapidly gaining wider acceptance in the U.S. The trend toward greater usage of resistive devices can be attributed to decreasing sensor and associated instrument costs. Table 2-1 shows a four-year market forecast for the five basic industrial sensors described in this article, both active and passive types, and their useful operating range. Each of these sensors has limitations, some of which can be alleviated by properly interfacing the sensor to its readout device and selecting the appropriate signal handling circuitry.

THERMOCOUPLE REFERENCE OPTIONS

Temperature measurements are commonly made with thermocouples which consist of two dissimilar homogeneous metal or alloy wires fastened together at one end to form a sensing junction and at the other end to form a reference junction. As shown in Fig. 2-1, temperature readout is obtained by connecting the two junctions in series across the input of a voltmeter. The voltage applied to the meter represents the temperature difference which exists between the two junctions. Since in most cases thermocouple wire resistance is only a few ohms, each TC junction can be modeled as a perfect voltage source whose output is proportional to temperature. The amount of current flow through the meter's input resistance depends strictly on the combination of thermocouple materials, the temperature difference between both junctions and loop resistance.

To facilitate measurement and calibration, the temperature of the reference junction is usually held constant. The voltage applied to the meter represents the temperature difference in relation to the reference junction. Most tabulations correlate this value to the triple point of water or 0 °C. In practice, one of the thermocouple junctions may actually be immersed in a saturated ice bath to obtain a stable reference point with an accuracy greater than 0.1 °F.

All thermocouple wiring between the point of measurement and the indicator should be of the same material, but less expensive alloy extension wires, which approximate the characteristics of the TC material, can be used to reduce costs.

If the wires which connect to the meter consist of copper material that closely matches the thermal characteristics of the input terminal material of a precision voltmeter whose input resistance is many decades greater than the TC loop

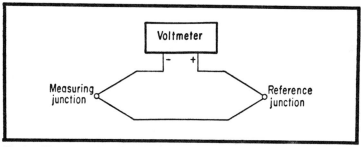

Fig. 2-1. The meter reading or temperature readout is proportional to the difference in temperature between the measuring and reference junctions. The reference junction temperature, usually held constant, can be immersed in a saturated ice bath to obtain an accurate and stable reference point.

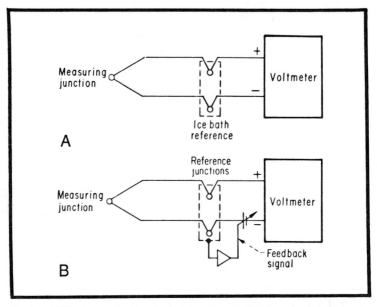

Fig. 2-2. A) The measuring junction can be referenced to ice point, even though the thermal characteristics of the TC and the meter input leads differ, by holding both reference junctions at the same temperature and using the same material in both of its legs. B) In this scheme, both reference junctions float at a precisely sensed temperature. Reference temperature changes are compensated by a feedback amplifier which adjusts a variable series voltage to maintain a net constant offset with respect to ice point.

resistance, the thermocouple acts as an almost perfect voltage source, and the loop voltage is proportional to the temperature difference.

If the wires which connect to the meter are not of the same material as the couple, the TC may still be referenced to the ice point, as shown in Fig. 2-2A. Any material may be used on the instrument side of the reference junctions, provided the same material is used in both legs and both junctions are held at the same temperature. When these conditions are met, the voltmeter reads the correct ice-point reference voltage.

Two further refinements can be made to simplify the measurement. One of the reference TCs may be held at an arbitrary temperature other than the ice point. Since a TC which undergoes temperature changes produces a predictable change in voltage output, a constant voltage source may be inserted in series with the constant temperature TC to exactly cancel the difference in output between it and the ice-point reference TC.

In another scheme that eliminates the need for an ice bath, Fig. 2-2B, both reference junctions "float" at a precisely sensed temperature. A feed-back signal from the sensing element causes an offset voltage in series with the negative lead to track changes in the reference temperature, thus producing a net constant offset with respect to the ice point. This feedback signal could also be converted to digital form to correct offset in the indicator itself before temperature is displayed. Such floating reference junctions are commonly used as "low-lead compensators" on modern temperature indicators. This method reduces the ambient temperature effects to approximately 0.01 °C error for each degree change in the reference temperature that normally tracks ambient.

PARASITIC EFFECTS

Figure 2-3 shows a complete thermal circuit for a typical TC indicator, where the final connection is made to the silicon substrate of an amplifier chip which increases the input voltage level to make stray thermocouple effects negligible. Although

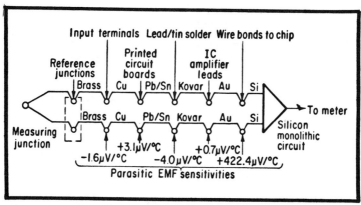

Fig. 2-3. Each of the stray thermal or parasitic emf's which effect TC output before it is amplified for meter display are shown here. These emf's introduce the largest error into TC measurements because TCs generate low-level output levels.

these stray thermal emf's affect all types of input sensors, they introduce the largest errors for TC measurements because TCs generate low-level signals in the millivolt range; a 1°C temperature change typically produces a 40 μV change in output. At each connection point in the loop, the parasitic thermal effects are associated in pairs. The effect of opposing parasitic pairs can be canceled by close thermal coupling. Note that some of the parasitics are extremely sensitive to temperature difference, especially at gold-silicon junctions. Thus, extremely tight thermal coupling between connections is required to assure accurate temperature readings. Low-level amplifiers may have to be used. Otherwise, comparatively large voltage offsets in the 10 − 100 μV range, associated with thermal time constants, may appear during the intitial instrument warm-up period. In addition, rezeroing schemes minimize amplifier offsets and slow-drift terms.

The proper application of thermocouples involves more than simply selecting a device that generates output greater than the expected parasitics. Consideration must also be given to useful operating range, corrosion resistance and stability. In addition to the standard thermocouples, there are 50 to 100 special-purpose TCs currently available which have a wide range of thermal characteristics.

A major limitation of conventional TC temperature sensors is initial accuracy. Even special accuracy material produces large errors compared with the capabilities and requirements of start-of-the-art instrumentation. However, in many applications, repeatability and long-term stability are more important than absolute accuracy.

COMPENSATING OFFSETS IN RTDS

Resistance temperature detectors (RTDs) utilize the temperature coefficient of resistance as the basis for measuring temperature. Various circuits can generate an output signal for readout as a function of resistance change, including classical bridge configurations. Most metals used to form RTD sensors exhibit a positive temperature coefficient of resistance, a relatively stable function as compared to the output of thermocouples.

One of the main disadvantages of RTD measurements has historically been the high cost of sensors, but less expensive ones are becoming available. The popularity of RTDs has increased because their lead wires cost much less than the extension wires normally required for TCs. Improved platinum sensors and other materials such as nickel are becoming more widely used.

The metals commonly used for RTDs are highly conductive; typical resistance values range from 10 Ω to 1 kΩ. Therefore, the series lead-wire resistance can introduce significant error into the measurement. It is not unusual for the parasitic voltage drop in the lead wires to exceed the drop in the RTD itself. Due to their relative insensitivity to parasitic thermal emf's, platinum or nickel RTDs can operate quite satisfactorily with standard copper-plated extension wires.

Figure 2-4A illustrates the conventional Kelvin four-wire connection often used to eliminate such errors. In this scheme a constant current generator creates an accurate voltage drop across the sensor which is read by a voltmeter having a high input resistance. If both the current source and the voltmeter were ideal devices, no parasitic offset could occur in series with the voltmeter leads, since current cannot flow through infinite input resistance. In addition, the drop that occurs in series with the current source would cause no error because it does not alter the current flowing through the source.

Since the voltmeter and the current source are not ideal devices, some error current flows into the voltmeter, usually through the negative lead. The non-ideal behavior of these devices results from imperfections in the voltmeter's input amplifier and from imbalances in the current source. These error terms can usually be considered negligible since the cable resistances normally encountered are less than 100 Ω per lead.

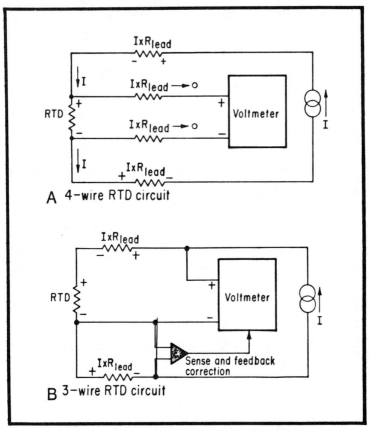

Fig. 2-4. A) This circuit eliminates errors resulting from parasitic voltage drops ($I \times R_{lead}$) in the lead wires which exceed the RTD voltage drop. The constant current source and a high-input resistance voltmeter minimize errors by developing an accurate voltage drop across the sensor and drawing minimal current from the source. B) This three-wire configuration, which reduces wiring costs, compensates for current imbalance in the sensor leads by sensing the voltage drop across the negative lead and feeding back a correction signal to the voltmeter.

Sensor costs can be minimized by using a three-wire RTD configuration to reduce the amount of copper lead wire required for measurement. This scheme assumes that the resistance of the lead wires has been matched to within 1 percent. As shown in Fig. 2-4B, the voltage drop in the negative lead can be measured and applied to the instrument as a feedback correction voltage. This reference voltage, which determines the amount of compensation required for the positive RTD lead, can be applied as an analog signal or changed to digital form by an A/D converter.

Span errors are commonly minimized with ratio techniques which derive the reference current from the same source as the A/D converter's reference voltage. Thus, errors resulting from drifts in the reference current cancel almost completely.

The output level of a typical RTD is higher than the output of a thermocouple. Since measurement is based on the increase of resistance with temperature, excitation current must be supplied to the sensor so that a measurable voltage output can be generated. However, the reference current cannot be arbitrarily increased to obtain a larger output since the RTD would self-heat and introduce serious errors into the measurement.

For a typical 100 Ω sensor, a 1 mA current source dissipates 100 μW, a negligible amount of power in most applications. At this current, a 100 mV drop is developed across the sensor at the ice point. In bridge configurations this drop is often compensated with series offset ahead of the voltmeter. Since the sensitivity of the sensor near the ice point is approximately +0.39 percent/°C, an offset of 390 μV/°C at the ice point is developed; this level is approximately an order of magnitude greater than most thermocouple outputs.

SOLID-STATE SENSORS

Thermistors are semiconductor resistors which exhibit a highly negative temperature coefficient. Although they have a more limited temperature range than RTDs, their extreme sensitivity to temperature changes makes them especially useful in high gain loops. Although their temperature vs resistance curve is highly nonlinear, thermistors can be interchanged in some ranges while maintaining an accuracy of 0.1 °F. Measurement methods for thermistors are the same as for RTDs, except that two-wire connections are commonly used (without lead compensation). Since the ice point resistance of these devices varies from about 1Ω to 1 MΩ, lead resistance tends to be relatively unimportant, especially at the higher thermistor values.

One disadvantage of thermistors is that they demand a wide dynamic range in the measurement system. Even though digital linearization may be used, it is difficult to display 0.1 °F outputs over more than three decades of resistance with a 16-bit A/D converter. Either the range of application must be limited or a swamping resistor must be connected in parallel with the thermistor to limit dynamic changes for easier linearization.

Semiconductor junction devices, both diodes and transistors, can also serve as temperature sensors, since they are highly sensitive and linear over a 200 °C range extending from below 0 °C to above 100 °C. PN junction diodes exhibit an almost linear voltage-temperature characteristic when forward current I_f, is held constant. In transistor sensors, if collector current I_c is held constant, then the base bias voltage V_{be} becomes a nearly linear function of temperature, over small ranges.

Three basic types of circuits have been developed to linearize temperature indications provided by transistor sensors. The first type holds collector current constant. However, this technique has some inherent nonlinearities. The second type uses feedback compensation to control collector current and cancel out nonlinearities. The third type utilizes either a step in collector current or a dual transistor which has different collector currents in each half. The difference in V_{oe} at the two collector currents is linear in temperature within fractions of a degree over a 150 °C span.

While it is more difficult to construct, the second circuit can provide better high-temperature linearity up to the 200 °C operating limit of the sensing transistor. The last two techniques provide better linearity than the constant current circuits.

RADIATION PYROMETERS

Radiation pyrometry, an indirect noncontact technique, permits a target's temperature to be inferred by measuring its radiated energy. This method has the following advantages: 1) it can measure temperatures above 3,000 °F; 2) it neither damages the measured object nor is itself harmed by the object; and 3) it can monitor moving targets and large surfaces. Either spectrally selective techniques or total radiation can be used for temperature measurement. Brightness or optical pyrometry is the most common form of the latter technique.

The application of broadband radiation pyrometers is complicated by the dependence of source radiant energy on both emissivity, a property of the radiating substance, and the physical peculiarities of the source surface. The combination of these effects is referred to as emittance. These difficulties are largely circumvented in practice by process calibration using a secondary measurement standard which has known characteristics.

Bandpass pyrometers find application where a narrow spectral range of emission is of interest. For example, the temperature of molten glass, which has a narrow (4.8 to 5.2 μm) range of emittance, can best be measured with an instrument which will pass only this wavelength band. Interference can be reduced by choosing passbands to reject absorption lines of intervening gases and vapors.

Ratio pyrometers measure energy in two separate bands and compute the ratio to infer temperature. This technique is especially useful for rejecting interference along the sighting path, when such interference affects energy in both bands equally. The most common optical pyrometers require an operator to match the brightness of reference source and target as seen through a color filter. Optical pyrometers are less adaptable to process control than radiation pyrometers.

By far the most common radiation device is the broad-band or total radiation pyrometer. By using suitable optics, target temperatures between 500 °F and 3,400 °F can be measured. Specially constructed units can measure up to 7,000 °F.

Total radiation pyrometers focus incident radiation on a thermocouple or thermopile detector. Thermopiles, which consist of a number of series-connected TCs that have compensated cold junctions, may take 7s to respond to thermal radiation. Although individual TCs respond faster, they produce low output levels.

MULTIPLEXING SENSOR DATA

Many applications require that a large number of temperature measurements be made at a minimum system cost per measurement. In such cases analog multiplexers are often used to connect several sensors into a single A/D converter which feeds a data system. This configuration allows the overhead costs of the measurement system to be shared among all of the sensors, thus reducing the cost per point.

There are several fundamental requirements imposed upon such multiplexers, especially since they must often operate at low signal levels. Multiplexers usually have differential inputs where pairs of sensor leads, rather than only one lead, are switched to avoid serious ground loop errors. Furthermore, the multiplexer must switch the voltage level of the input sensor with respect to ground, i.e., the common-mode voltage. Common-mode voltages may reach several hundred volts in some applications.

Figure 2-5 illustrates a simple two-channel multiplexer system and common-mode sources. Although thermocouples are shown supplying the input voltages, the basic principles apply to all sensors. The common-mode voltages are represented by AC and DC components, V_{cm}, in series with common-mode source impedance, Z_{cm}. Clearly, if a single-ended multiplexer were used, the lower poles of S1 and S2

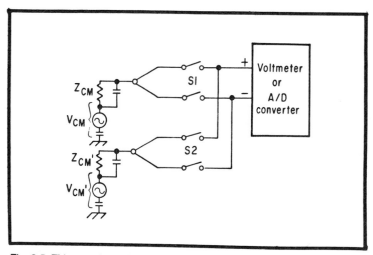

Fig. 2-5. This two-channel multiplexer switches both leads of each TC into a single voltmeter or A/D converter to avid ground loop errors. Since the common-mode voltage (V_{cm}), which may exist between the floating sensor output and earth ground, can cause parasitic currents to flow in the negative lead, source impedance Z_{cm} must be maximized or V_{cm} minimized. Differential amplifiers are used to handle floating inputs in most industrial temperature indicators.

would be shorted, causing both common-mode sources, which have unequal outputs, to drive parasitic currents through the negative leads.

Only 1 mA of parasitic current is required to produce a 1 μV error in 1 Ω of lead resistance. This calculation suggests that either Z_{cm} must be very large (approachable with floating, shielded TCs) or that V_{cm} must be very small (approachable with earth-grounded sensors). However, these limiting cases are rarely achieved. Floating sensors invariably pick up enough AC common-mode voltage through stray capacity to cause error current; grounded thermocouples can rarely be connected to identical earth potentials. Thus, differential multiplexers are required in almost all nonlaboratory environments.

Multiplexers must also withstand breakdown voltages applied to their switches. Breakdown potential is the difference in common-mode voltage V_{cm} imposed across a single pole pair when an adjacent point is closed. For breakdown potentials above 40 V, electromechanical relays are required;

below this level solid-state switches can be used. Cradle or crossbar relays provide the best performance, although reed relays offer economy.

A fundamental problem with relays in low-level measurements is that they require ferromagnetic pole or switch material which produces parasitic emf's. In large relays, the magnetic parts are separated from the contacts so that no parasitic thermal emf's can occur. However, the mechanical design of such relays usually increases the coil power requirement per point. Reed relays, representing an alternative switching method. are now available to match the low thermal offset characteristics of larger relays. Close thermal coupling of parasitic TC effects allows thermal offsets to be held below 1 μV.

Where high common-mode potentials do not represent a major problem, solid-state multiplexers may be used, and are significantly more reliable than relays. State-of-the-art LSI devices can multiplex microvolt-level signals while consuming only a small amount of power. For example, a 16-pin IC multiplexer based on dielectrically-isolated CMOS technology consumes only 1 mW per point, as opposed to 150 mW for a typical low-thermal, reed-relay pair.

So far only differential or two-lead multiplexers have been discussed. Clearly, more switches per point are necessary for three and four-wire connections, such as used with RTDs and thermistors. Thus temperature measuring systems which include resistive sensors cost more than those that receive inputs from TCs. A cost trade-off can be made, however, since lower performance multiplexers may be used to handle the higher level outputs of these sensors.

PART II: ENHANCING THE MAN/MACHINE INTERFACE

Maximizing the accuracy of temperature measurements may involve using several noise reduction techniques, since various sources can affect signal quality. The most frequently encountered form of noise is derived from common mode sources. For purposes of analysis, common mode noise may always be converted to an equivalent normal mode noise. Normal mode noise, developed in series with the sensor input, may also be present in the system. But because of the low sensor impedance of thermocouples and RTDs, it is usually less of a problem.

The main sources of common mode error are shown in Fig. 2-6A. Normally, the positive input lead of the A/D converter presents a higher input impedance to V_{CM} than the negative input lead, which is also typically power supply ground. This higher impedance consists of filter impedance in series with the high lead and a small amount of stray capacitance to earth.

Consequently, more input current, which flows through the negative lead, causes noise to appear across the lead impedance in series with the input signal. If the A/D converter has sufficient normal mode rejection at the noise frequency, the resulting error may be negligible. However, the amount of rejection at frequencies below the cutoff of the input filter may be inadequate and other rejection schemes must be used.

REJECTING COMMON MODE NOISE

Guarding, a common method to improve common mode rejection (CMR), has two forms: active and passive. In both forms, negative and positive leads are shielded from the stray case capacity to minimize the parasitic currents flowing through the lead resistance. As shown in Fig. 2-6B, when the shield is tied to the common mode source, an alternate path is provided for the current resulting from the stray capacitance to earth ground. Since less parasitic current flows through the lead resistance, less noise voltage develops across the thermocouple leads. Unfortunately, if a multiplexer is used, a separate switch pole is required on each thermocouple to switch the common mode voltages. The additional hardware for guard switching can increase the cost of the multiplexer by 50 percent.

An alternate scheme which requires no guard switching can be used. As shown in Fig. 2-6C, an earth-referenced or guard amplifier can be used to measure the analog ground potential which, except at very high signal frequencies, is almost equal to V_{CM}. The guard amplifier, whose output is automatically switched by a two-pole multiplexer, thus tracks the input V_{CM}. The output of the guard amplifier drives the guard shield.

In some cases, neither of the above schemes provides adequate rejection of input common mode noise at higher frequencies. For example, various stray capacitance paths within the instrument radiate high frequency signals directly into sensitive preamp nodes. Most solid-state amplifiers rectify or detect signal frequencies above 10 to 100 kHz and can produce millivolts of offset at frequencies up to 1 GHz.

High-level signals at such frequencies are becoming more common in industrial environments due to faster logic, fast thyristor controllers, and high frequency oscillators. Noise spikes of various kinds up to several hundred volts in amplitude, common on local power line circuits, enter the instrument as high-frequency common mode errors. In many cases, connecting high-frequency bypass capacitors to earth ground at the instrument input terminals more effectively reduces noise than common mode shields. A combination of the schemes shown in Fig. 2-6 (B and C) provides good CMR to at least several hundred megahertz.

Another frequently overlooked source of noise is the common mode exit current of the instrument itself. This current, which flows through high source impedance, exits through the negative terminal of the instrument and through the low-lead source impedance as common mode current. This noise current usually results from inadequate power supply transformer shielding or the close proximity of fast logic circuits to earth ground.

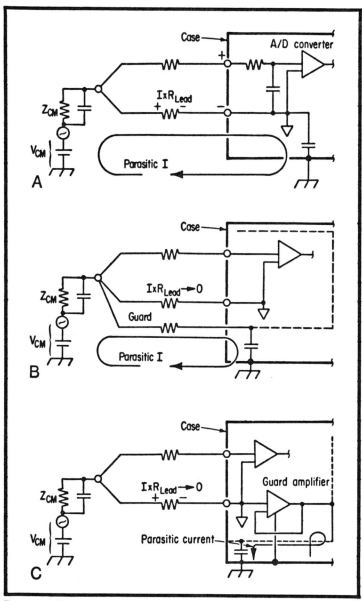

Fig. 2-6. A) Parasitic common mode current which flows through the low impedance negative lead causes temperature measurement error by generating noise. B) If the A/D converter, shown in A, cannot reject noise passed by the RC input filter, passive guarding can be used to minimize parasitic currents. Tieing the A/D converter shield to the common mode source provides an alternate current path which reduces the flow through the lead resistance. C) This active guarding scheme employs an earth-reference amplifier to track input Vcm and drive the guard shield.

Although the frequency of transformer noise is that of the power line (e.g., 60 Hz) and is rejected substantially by dual-slope integrating A/D conversion methods, it must still be kept low. Unless adequate poweer line RFI filtering is incorporated into the power supply, high levels of 60 Hz current tend to exit through the input sensor during power line spikes. Noise resulting from logic "pumping" into earth can be minimized by careful ground plane shielding. Furthermore, if switching regulator or DC/DC converter supplies are used, exit spikes can be very large unless careful shielding is employed. In general, less than 100 nA peak-to-peak of 60 Hz "pumpout" can be achieved.

Dual slope converters, if used at multiples of the power line frequency, provide an excellent economical method of rejecting line related noise. Integrator A/D schemes such as dual slope must have a maximum speed of 60 readings per second or less to reject 60 Hz noise. Faster conversion methods do not ha· e the inherent noise rejection capability discussed above.

LINEARIZATION METHODS

Most thermocouples are nonlinear enough to require some compensation to achieve 1 °C accuracy. Thermocouple linearization curves for 0.1 °F accuracy typically involve at least a seventh-order poly-nominal, while platinum RTDs involve only a second-order polynomial. Linearization may be analog, such as a diode breaking, nonlinear gain stage, or it may be accomplished digitally with hardware or with software.

Most conventional linearization techniques are digital. Digital techniques have a cost and accuracy advantage over analog methods. In either case, knowledge of the nonlinearity characteristics of the sensors is required. Standard tables and polynomials are available from the NBS for most major types of TCs and platinum RTDs; such standards as German DIN documents apply to European sensors. Although thermistor standards are not yet established, data which can be used to define nonlinearities can be obtained from sensor vendors.

In nonprocessor-oriented instruments, a binary rate multiplier (BRM) and read-only memory (ROM) table look-up procedure may be used with a dual slope A/D converter to linearize the measurement. During the reference integration period, the BRM varies the counter frequency in a piecewise linear segment fashion so that A/D counts accumulate in a precise, but nonlinear fashion. Since the count is directly proportional to the linearized input voltage, it accurately represents the temperature measured. If a processor is available for linearization, the problem is conceptually simpler. Where the polynomial curve fit for the sensor is known, the processor needs only to evaluate the polynomial at any count received from the A/D converter. In most cases, polynomial evaluation represents an unnecessary workload for the processor, even though a minimum of stored data is required (N polynomial coefficients). The software evaluation is often excessive and inefficient.

A faster method which does not require a large number of high-precision multiplications consists of a table look-up for a piecewise linear error correction to any nonlinearized A/D converter output count. This method can be very fast and does not burden the processor with numerous calculations. Other digital linearization algorithms undoubtedly will be developed that are especially suited to the present generation of ultra-low cost microprocessor components.

SIMPLIFIED MAN/MACHINE COMMUNICATIONS

The operator interface of a data acquisition system has often been considered secondary to its technical aspects and specifications. However, the ability of an instrument system to provide meaningful, timely communication with a human operator is becoming of primary importance.

We are well aware of how easy it is to overload a human decision-maker with data. The reams of computer printout often produced by EDP operations serve as a harsh reminder of the confusion between data and information. The manner in which information is displayed can often make the difference between effective utilization by decision-makers and neglect or, worse, misinterpretation.

Once temperature sensor signals have been converted to digital form, they can be read out in various ways. In the simplest case, the data can be displayed as a temperature readout on a panel indicator. For such applications as monitoring a single process temperature loop, an accurate, readable indicator is adequate. More complicated processes which often have multiple interacting thermal variables involve a different kind of operator interface. In such processes, the operator needs additional information, including identification of the thermal variable, the time of reading, and limit or alarm conditions.

In some applications, the relationship between thermal variables may be as important as the levels themselves. For example, when the temperature levels of input variables at the time of reaction vessel alarm condition must be known, indicator displays alone are inadequate, especially where many variables are involved. Such applications have been traditionally served by multipoint recorders. Digital data loggers have recently been applied as well, but they have not been cost-competitive with multipoint recorders.

Multipoint recorders graphically present the magnitude and time of reading of each variable. The juxtaposition of multiple variables on graduated chart paper in multipoint recorders, both circular and strip types, requires visual interpretation to determine relationships of variables to limits, the time of measurement, and the identification of the channel itself. Physical limitations place an upper limit on the number of channels which can be effectively treated in this manner.

ENTER THE DIGITAL DATA LOGGER

The digital data logger can handle very large numbers of variables, remove ambiguities of interpretation and provide for sophisticated alarm strategies. Smart data loggers implemented with microprocessors have brought the level of sophistication, vis-a-vis alarm strategies, up to a point previously attainable only with central computer control.

The strip chart data logger overcomes the difficulties associated with numerous variables by printing point identification numbers along with data and time-of-day with each scan. Numeric data presentations eliminate possible ambiguities associated with analog records. Data presented in a fixed format and data fields have implicit positional significance, as chown in Fig. 2-7.

Although the fixed data format is readily mastered, the operator must nevertheless become acquainted with it before he can interpret it. Although the limits applied to a particular point may be determined by a simple interrogation, they are still a step removed from the operator's attention. However,

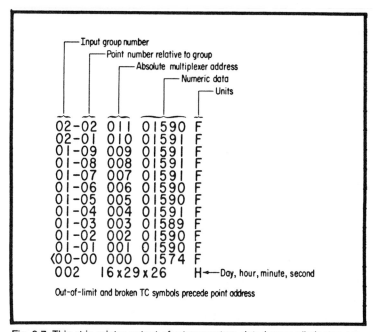

Fig. 2-7. This strip printer output of a temperature data logger eliminates ambiguities associated with analog records by presenting numeric information in a fixed format. The position of the data fields shown here has implicit significance.

limit conditions (not to be confused with limit values) are explicitly indicated by the strip printer log.

Point identification for multipoint recorders is implicit; that is, the nth pen position and nth graph correspond to a particular sensor input. Most data loggers record the point number along with point data. Points may be serially ordered 0 through 1,000, or they may follow a hierarchy of group and point number.

Inputs to multipoint data scanners and loggers are configured in hardware to be as general-purpose as possible. The unique attributes of an input channel are best handled in firmware. For example, J-, K- and T-type thermocouples may be mixed on a multiplexer board and their particular linearization curves can be stored in ROM. While input conditioning for RTDs is physically different than for thermocouples (which require different front-end hardware), RTD technologies can themselves be mixed and the unique characteristics handled in firmware. Both technologies can be included with the same system.

It is desirable in some applications to record the unlinearized input potential. The dynamic range of conversion can be preserved over a span of 10 μV to 10 V or more by breaking the span into ranges. Input channels, or points, may then be assigned to particular gain ranges. Where autoranging is available, gain ranges can be selected automatically by the instrument.

Temperature measurements may be linearized in °F and converted to °C, or vice versa. Fahrenheit linearization yields better accuracy since it has higher resolution. Gain ranges, linearization curves, display units, and offsets can all be assigned on a per-point basis with modern data logging equipment.

LIMITS, SCAN DISCIPLINES FOR LOGGERS

Limit comparison is fundamental to any alarm system. Depending on the application, it may be desirable to limit a variable from above (a high limit) or from below (a low limit) or to assign priorities to such limits in a given direction and to distinguish limit directions. For example, calling the operator's attention to a point temperature above 500 °F or below 490 °F may be desired. If, however, the temperature excursion were to go above a second high limit of 510 °F, it may be desirable to sound a general alarm or to take immediate remedial action.

Up to four alarm limits may be assigned and individually alarmed per input point on some modern microprocessor-controlled data loggers. It is advantageous to be able to both assign particular alarms (solid-state relay contact closures for external control) to any limit condition on any point, and to enable or disable these alarms at will. When establishing an alarm limit philosophy, it may be expedient to incorporate one of the following conditions in the data logger: alarm condition terminated either by the points returning within limit, or by an operator's acknowledgment.

In general, the order in which points are scanned is less important than making sure that all points are examined in a timely manner. For example, it may be desirable to examine points 49 through 100 every 5 minutes and to check points 1 through 48 and point 50 once a minute. Additionally, it is also convenient to be able to examine any point or group of points at any time on demand, without disturbing the specified scan discipline. The point-skip approach is another useful modification of any scan discipline. It may happen, for example, that a particular machine must go through a shut-down sequence for maintenance, but that other machines in the process remain on line. In this application, it is desirable to skip the points associated with the disabled machine without disturbing the established scan discipline.

Logging refers to the act of storing data on printed paper strips, punched paper tape, magnetic tape or other recording media. The examples which follow illustrate the kind of logging strategies made possible by flexible firmware control in modern data acquisition and logging systems.

To continually log all data that is scanned for alarm conditions can produce an immense amount of data. To limit the amount of data, all data could be logged only whenever an alarm condition occurs. Or, it may be desirable to log only alarm condition data while continuing to scan all points as determined by the scan discipline imposed.

Some logging schemes may not require alarming functions, but only a temperature profile recorded at given intervals as required by environmental regulatory agencies. Another option is for a log of all points currently in alarm, on demand, without disturbing the established logging discipline. Thus, for example, an operator may request an alarm demand log on a teletype console in the control room without disturbing the data being logged on magnetic tape.

A feature known as nuisance delay allows an alarm to be conditioned on both limits and the number of occurrences of the out-of-limit condition. For example, point X is scanned every minute, but it causes an alarm only after the third out-of-limit condition. In this example, the consequent delay would be 3 minutes. Out-of-limit fluctuations for less than 3 minutes are discriminated against.

PERIPHERAL DEVICES FOR DATA ACQUISITION

Formatted CRT displays offer solutions to most of the interface problems discussed previously. Figure 2-8 illustrates one format where column headings indicate point numbers, variable values and alarm limits. Temperature values are updated in place on the CRT display and out-of-limit conditions are displayed in a reversed field block (dark characters on a light background). Skipped points show a point number, but no value or limits are listed.

Limits and skip conditions can be edited without disturbing the alarm scanning. It is also possible to utilize reserved

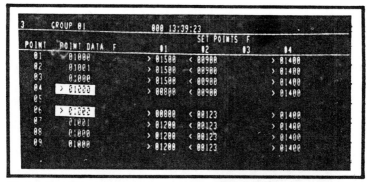

Fig. 2-8. This CRT display shows a list of temperature sensor points which have been formatted to provide meaningful, timely information for the operator. Column headings indicate points, values and limits; values are updated in place; and limits and skip conditions can be edited without disturbing alarm scanning.

fields on the CRT form to prevent editing of particular point limits. Cursor control also allows for a character edit feature where the display can be edited without ever leaving the forms mode. That is, the cursor can be positioned under the character to be changed and the new value entered. Flashing fields can be utilized to indicate broken thermocouples.

The only remaining criticism of the above display is that the operator must still associate an abstract point number with a process variable. Prompts could be issued to give a brief narrative point definition, but they would interrupt the useful display format.

Graphic CRT terminals have the potential of solving this last problem, although at the expense of the compactness offered by the formatted CRT display. Process piping diagrams can be displayed with temperature values shown next to the graphic location of the temperature sensors. Thus, the CRT screen can provide a snapshot of a graphical control board. The obvious advantage of the CRT is that attention can be directed at any part of the process flow and changes can be made in firmware rather than on an actual control panel.

The data management environment of a data acquisition and logging system determines, to a large degree, the peripheral devices which should be used with that system.

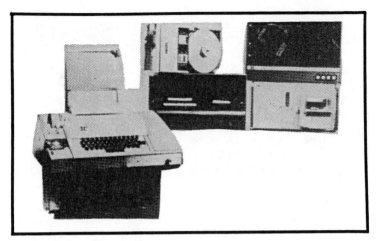

Fig. 2-9. Some of the standard peripheral devices which can be interfaced to modern data acquisition and logging systems include a Teletype, a magnetic tape logger and a paper tape unit. Such peripheral devices are selected on the basis of the data management environment, record lengths, unattended service interval and operator preference.

Length of record, unattended service interval and operator preference also enter into the selection process. Figure 2-9 illustrates some of the standard peripheral devices which can be interfaced to modern data acquisition and logging systems.

For example, it may be necessary that a single system provide both real-time control room monitoring and data recording for off-line process analysis. The data processing center may be able to process data stored on both punched paper tape and nine-track, IBM-compatible-format magnetic tape. It may also be desirable to service the recording device only at shift changes while recording over 100 points of data every 5 minutes. In this application, magnetic tape would be the logical choice for the logging device.

The capability of a CRT keyboard display terminal to obtain rapid alarm and demand logs makes it a desirable choice for control room applications.

Remotely locating a data logger where operation is unattended may make telecommunications desirable. Such data loggers can use modems to transmit control signals and data to extend their capability. On-site point data acquisition can be important for machine adjustment and maintenance, but using a large communications terminal such as an ASR-33 is undesirable in a machinery floor environment. Manual point entry with a portable local keyboard display unit, Fig. 2-10, can fill this need.

MICROPROCESSOR CONTROL

The previous discussions have assumed that editing of the scan disciplines, limits and alarm strategies can all be done without disturbing the initial alarm scanning modes. We have also indicated that demand logs can be requested by display terminals, including remote keyboard display units without disturbing the alarm scanning activities of the data acquisition system.

This capability is called foreground-background real-time operation. Since the microprocessor has a fast execution speed, there are periods when it idles after having computed all its scheduled activities or while it is waiting for slower hardware. These time periods, available for background processing, occur frequently enough to give the operator the

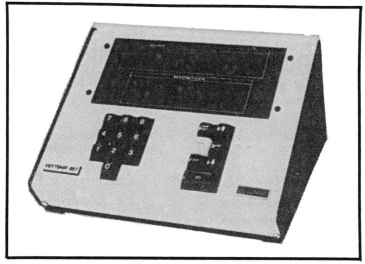

Fig. 2-10. This portable keyboard display unit, which facilitates manual point entry, can substitute for a large communications terminal.

impression that his keyboard is being serviced immediately. The real-time foreground-background mode is especially useful in real-time alarm scanning systems, but it is not widely available.

The CRT displays and remote keyboard displays discussed previously make use of distributed intelligence. These devices are themselves microprocessor-based with local firmware control. The advantage associated with the distribution of intelligence is that the operation of peripheral devices do not contribute to the workload of the data acquisition system controller. Furthermore, since these options can be programmed independently, their operational attributes can be changed without requiring reprogramming of the data acquisition system itself. Such changes are made at the factory. Operator control of all acquisition system and peripheral devices can be accomplished without special knowledge of computer programming languages.

BIBLIOGRAPHY

Temperature: Its Measurement and Control in Science and Industry, Parts 1-3, Vol. 4, H.H. Plumb, Ed., Instrument Society of America, Pittsburgh, 1972.

Verster, T.C., "The Silicon Transistor as a Temperature Sensor," *Temperature: Its Measurement and Control in Science and Industry*, Part 2, Vol. 4, Instrument Society of America, Pittsburgh, 1972.

Harrison, T.R., *Radiation Pyrometry and Its Underlying Principles of Radiation Heat Transfer*, John Wiley & Sons, New York, 1960.

Process Instruments and Controls Handbook, 2nd ed., D. M. Considine Ed., McGraw-Hill Book Co., New York, 1974, pp. 2-71 through 2-100.

DAVID M. MACKENZIE is Engineering Project Manager and WILLIAM E. KEHRET is Vice President of Engineering at Doric Scientific, San Diego. Article is based on a paper presented at the ISA Aerospace Industries and Test Measurement Divisions Symposium, San Diego, 1976.

Spectroscopic
Temperature Measurements

P.S. SCHMIDT

Recent advances in high temperature technology have brought a wave of new applications of hot ionized gases to chemical production, materials processing and biomedical engineering. Temperature measurements in ionized gases or plasma pose an especially difficult problem, however, because temperature levels are often many times higher than the vaporization temperatures of even refractory metals. Remote sensing of temperature is obviously required. Spectroscopy, the art of deducing the properties of substances by analyzing their electromagnetic spectra, lends itself to this type of measurement and to many other applications as well (Ref. 2.1, 2.2).

Several phenomena that occur at the atomic level provide a basis for determining the temperature of a hot gas. In molecular gases, the radiation produced by rotation and vibration of the molecules can be related to temperature. Broadening of spectral lines due to Doppler shifts resulting from random thermal motion of the emitting molecules can be measured when gas pressure and temperature are low enough that other broadening effects do not blank out the one desired. The absolute intensity of radiation resulting from electron

137

excitation can be measured, but the calibration of the detecting element, a spectrometer photomultiplier tube, must be absolute and numerous corrections are required to maintain accuracy.

A commonly used technique to determine the temperature of ionized gases employs the ratio of radiation intensities of different spectral lines resulting from the excitation of electrons in neutral atoms. This intensity ratio remains independent of the density of the gas and the number distribution of atoms in the various possible energy states (partition function). Since only a relative calibration of the photomultiplier tube is necessary, the two-line methods have a distinct practical advantage over absolute intensity methods. The primary uncertainty in the two-line technique stems from incomplete knowledge of the transition probabilities which will be described later.

READING SPECTRAL LINES

A typical system for making spectroscopic temperature measurements in homogeneous sources with two-line methods is shown schematically in Fig. 2-11. A mirror focuses two beams of light emitted by the plasma source on the entrance slit of the spectrometer. The lower concave mirror focuses the beams entering the spectrometer through the slit onto two plane gratings, which break them into wavelength components. The upper mirror then reflects the beams through the exit slit where a chopper alternately selects light from each emergent beam and flashes it upon the photomultiplier tube. The photomultiplier produces a minute electric current proportional to the intensity of each incident monochromatic beam; the amplifier conditions these currents for sample and hold and scaling circuits which determine the intensity ratio for plotting on a recorder. The photomultiplier tube must be calibrated to determine its relative sensitivity to various wavelengths with a light source having known spectral characteristics. To examine a particular region in the plasma,

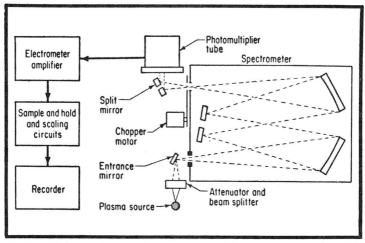

Fig. 2-11. This spectrometer system measures the temperature of the plasma source by determining the intensity ratio of two spectral lines: one emitted by ionized atoms and the other by neutral atoms. Due to broadening effects, it is necessary to scan on either side of a selected line and integrate the intensity over a finite wavelength band.

auxiliary optics would be required ahead of the spectrometer to permit fine resolution of the image projected on the entrance slit.

Due to broadening effects, the radiation emitted during an electronic transistion will not actually be limited to a discrete wavelength, but will be smeared over a narrow wavelength band with the intensity reaching a peak at the nominal wavelength of the line. To account for all the radiation belonging to a particular spectral line, it is necessary to scan the region on either side of the nominal wavelength and integrate, graphically or by other means, the intensity over the entire band. For this reason the experimenter must carefully choose the spectral lines to be employed in the measurement to avoid the overlap of wavelength bands.

The temperature measurement technique detailed here involves a variation of the two-line method in which one of the spectral lines is emitted by singly ionized atoms and the other by neutral atoms. This method is more accurate than the neutral-neutral technique in the high-temperature range (above 12,000 °K for argon, for example) due to the large difference in the energy levels of the two lines.

ELECTRONIC TRANSITIONS

When a gas has been energized to a very high temperature by combustion or by passage through a high-powered electric arc, for example, the electrons bound to each atom may become excited. During this energizing process, they may jump temporarily to energy levels above their normal ground state. Applying sufficient energy to the gas causes some electrons to actually free themselves from the rest of the atom; in this manner, plasma is produced. As these free or excited electrons return to their original energy states, the energy lost in the transition process is emitted as electromagnetic radiation at a discrete frequency.

The frequency of the radiation, v, may be determined from Planck's quantum theory as $E_n - E_m = hv$ where E_n and E_m are the energies of the upper and lower levels of the transition, respectively, and h is Planck's constant (6.626×10^{-34} J-s). The local intensity of the radiation, assuming no self-absorption by the gas (i.e., "optical thinness"), per unit area, time and solid angle from a source of depth d is:

$$I_{nm} = hv_1 A_{nm} n_n d/4\pi \qquad (2.1)$$

where A_{nm}, called the Einstein A coefficient or transition probability, represents the percentage of all electrons in state n which will spontaneously make the transition to state m in a unit of time. The value of A_{nm}, calculated from quantum theory for only a few isolated cases, usually must be obtained from experimental data. Factor n_n is the number density (number/unit volume) of atoms with electrons in the nth energy state. For a gas in local thermodynamic equilibrium, n_n follows a Boltzmann distribution:

$$n_n = n_j g_n/Z_j \, exp \, (-E_n/kT) \qquad (2.2)$$

where subscript j represents the particular ionization level of interest; i.e., $j = 0$ for neutral atoms, 1 for those singly

140

ionized, and so on, and Z_j is the partition function for the jth species, a known function of temperature from statistical mechanics. Factor g_n devotes the degeneracy or statistical weight of the n state, an integer which may be interpreted as the number of possible orientations of the electron's spin and orbital angular momentum and energy for a given electron state; this integer is a known invariant for a particular state. Factors k and T represent the Boltzman constant (1.38×10^{23} J/°K) and the temperature, respectively.

Substituting Equation 2.2 into Equation 2.1, radiation intensity I_{nm} can be expressed as:

$$I_{nm} = h v_1 A_{nm}\, n_j \quad \frac{g_n}{Z_j} \quad \frac{d}{4\pi} \quad exp\ (-E_n/kT) \qquad (2.3)$$

Direct applications of this equation are difficult because measuring the intensity of a single spectral line, as mentioned previously, requires an absolute detector calibration and frequent adjustments.

TWO-LINE RATIO METHODS

The difficulties may be circumvented by measuring a second neutral atom spectral line of frequency v_2 emitted in a transition from a state q to a lower level p. The intensity ratio of the two lines is:

$$I_r = \frac{I_{nm}}{I_{qp}} = KA\ exp\ \left[\frac{-(E_n - E_q)}{kT}\right] \qquad (2.4)$$

$$\text{where A} = \frac{v_1\, A_{nm}\, g_n}{v_2\, A_{qp}\, g_q}$$

and K is a calibration constant representing the ratio of the photomultiplier tube sensitivity at v_1 to its sensitivity at v_2. Note that the partition function and the number density have canceled in Equation 2.4, thus simplifying it considerably.

For high accuracy it is desirable to choose two spectral lines which will make I_r a strong function of T; this will be the case if E_n is chosen to be much greater than E_q. But for neutral atoms, the greatest attainable energy difference still yields a minimum uncertainty in temperature of 70 percent of argon in the 6,000 to 15,000 °K range. The accuracy may be improved somewhat by measuring several pairs of spectral lines and averaging the resulting temperatures. However this rather time-consuming technique is only partially successful in reducing the experimental uncertainty.

A larger energy difference can be obtained if one of the transitions, say from n to m, occurs in a singly ionized atom, and the other occurs in a neutral atom. The intensity ratio then becomes:

$$I_r = \frac{(I_{nm})_i}{(I_{qp})_a} = KA \quad \frac{Z_a n_i}{Z_i n_a} \quad exp \left[\frac{-(E_n - E_q)}{kT} \right] \qquad (2.5)$$

where the subscripts i and a represent the singly ionized and neutral atom species, respectively.

The number densities n of neutral atoms and ions, along with the corresponding partition functions Z have now been reintroduced in the equation; but these functins can be calculated from the Law of Mass Action applied to the ionization reaction (the Eggart-Saha equation) of argon, for example: $Ar \leftrightarrows Ar^+ + e^-$.

$$\frac{n_i n_e}{n_a} = 2 \left(\frac{2\pi \, m_e \, kT}{h^2} \right)^{1.5} \frac{Z_i}{Z_a} \quad exp \left[\frac{-(\epsilon_i - \epsilon_a)}{kT} \right] \qquad (2.6)$$

where n_e is the electron number density.

Neglecting the effect of lowering the ionization potential, the differences of energy levels $(\epsilon_i - \epsilon_a)$ is simply the ionization energy E_i. Substituting E_i into Equation 2.6 and rearranging:

$$\frac{Z_a \, n_i}{Z_i \, n_a} = 2 \left(\frac{2\pi \, m_e \, kT}{h^2} \right)^{1.5} \frac{exp\left[\frac{-E_i}{kT} \right]}{n_e} \qquad (2.7)$$

The expression for the relative intensity then becomes:

$$I_r = \frac{2K}{n_e} \left(\frac{2\pi \, m_e \, kT}{h^2} \right)^{1.5} A \, exp \left[\frac{-(E_n - E_q - E_i)}{kT} \right] \quad (2.8)$$

Note that for the ionized case, the numerator of the exponential is greatly increased by the addition of E_i. Substituting the appropriate values of the physical constants:

$$I_r = \frac{(4.82 \times 10^{15}) \, K}{A T^{1.5}} \quad (2.9)$$

where $B = \dfrac{E_n - E_q + E_i}{K}$

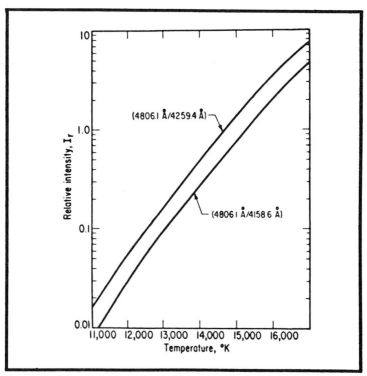

Fig. 2-12. For argon above 12,000 K, the relative intensity of two ion-neutral spectral line pairs is quite temperature-sensitive and relatively linear when plotted on semilog coordinates. Plots were calculated from reference data, assuming a pressure of one atmosphere.

Constants A and B are fixed for any selected pair of spectral lines. For argon, various investigators have determined the transition probabilities, statistical weights, and upper energy levels for a number of spectral lines (Ref. 2.3); and the electron number density as a function of temperature (Ref. 2.4).

Using these data, Incropera (Ref. 2.5) calculated the dependency of I_r on T for two ion-neutral line pairs, Fig. 2-12. Note that, for these line pairs, the relative intensity is extremely temperature-sensitive. Below about 12,000 °K, the ion-neutral two-line method loses its effectiveness due to the low degree of ionization in the gas.

When measuring temperatures in an axisymmetric source, such as a plasma jet, an additional step is necessary to reduce the experimental data. Figure 2-13 illustrates that at any lateral location, the spectometer slit "sees" not only the emission from the gas corresponding to the radial position of the slit, but also the emission from the other temperature regions or isotherms interesected by the slit's field of view.

It is therefore necessary to convert the measured lateral intensity distribution across the source to a true radial intensity distribution before applying Equation 2.9. A method to

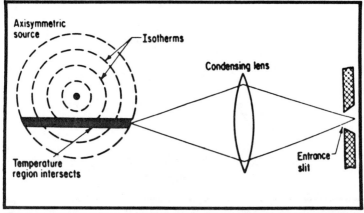

Fig. 2-13. Emissions from undesired temperature regions in the plasma source intersect the entrance slit of the spectrometer and should be compensated by converting lateral intensity distribution across the source to true radial intensity.

accomplish this conversion, devised by Nestor and Olsen (Ref. 2.6), essentially involves measuring the total emitted radiation starting at the outermost radial position, and successively subtracting out the volume-weighted contributions of each "piece" as the slit's field of view advances toward the center. Due to geometric considerations, the Nestor and Olsen conversion is somewhat involved algebracially, but it may easily be incorporated into a data reduction algorithm, so that radial temperature profiles can be computed directly from lateral intensity distribution measurements.

REFERENCES

2.1 Griem, H. R., *Plasma Spectroscopy*, McGraw-Hill, New York, 1964.

2.2 Dickerman, P. J., ed., *Optical Spectrometric Measurements of High Temperatures*, University of Chicago Press, Chicago, 1961.

2.3 Olsen, H. N., "Measurement of Argon Transition Probabilities Using the Thermal Arc Plasma as a Radiation Source," *Journal of Quantitative Spectroscopy and Radiative Transfer*, 3, 1963, pp. 59-76.

2.4 Drellishak, K. S., et al, "Partition Functions and Thermodynamic Properties of Argon Plasma," AEDC-TDR-t3-146, August 1963.

2.5 Incropera, F. P., "Temperature Measurement and Internal Flow Heat Transfer Anaysis for Partially Ionized Argon," TR Number SU 247-11, Stanford University, August 1966.

2.6 Nestor, O. H. and Olsen, H. N., "Numerical Methods for Reducing Line and Surface Probe Data, *Siam Review*, Vol. 2, No. 3, July 1960.

PHILIP S. SCHMIDT is an Associate Professor in the Department of Mechanical Engineering at the University of Texas, Austin.

Humidity Measurements

P.R. WIEDERHOLD

PART I: PSYCHROMETERS AND PERCENT RH SENSORS

Humidity can be expressed in a variety of different forms: wet bulb temperature, percent relative humidity, vapor pressure, mixing ratio, dew/frost point, grains per pound, parts per million and others. These parameters can be measured by a number of different instruments, each capable of accurate measurement under certain conditions and within specific limitations.

An instrument engineer faced with a moisture measuring problem not only must decide which humidity parameter he should measure, but he must also select an instrument which is best suited for that measurement in his particular environment. To make the problem more complex, the form in which humidity data is needed may differ from the form provided by the most appropriate sensor.

In meteorology, the public wants to know the relative humidity on the evening newscast. Meteorologists calculate the percent relative humidity with a circular sliderule, using data gathered by a saturated salt dew sensor. In their particular environment, a dew point sensor provides more accuracy than an instrument which measures percent relative humidity directly.

This article deals with the various types of humidity measurements and the sensors and instrumentation commonly used in industrial automated systems. Only those sensors in wide use and for which error analyses have been published by the National Bureau of Standards are considered. Particular attention is given to the range, response and accuracy which can be expected from each type of instrument.

DEFINITION OF HUMIDITY

Unless one is routinely working with humidity measurements, there is a tendency to forget that humidity is water gas, behaving in accordance with the ideal gas laws. One of the easiest ways to put humidity in its proper perspective is through application of Dalton's law of partial pressures to the most commonly encountered gas—air.

Dry air is a mixture of nitrogen (78 percent) and oxygen (21 percent), with carbon dioxide and other minor constituents making up the remaining 1 percent. The percentage of water gas in air is represented by the humidity, and this percentage increases from zero percent in dry air to a value of 4.5 percent (by volume) at 100 percent relative humidity and a temperature of 90 °F.

Dalton's law states that the total pressure, p_m, exerted by a mixture of gases or vapors is the sum of the pressures of each gas if it were to occupy the same volume by itself. The pressure of each individual gas is called its partial pressure. The total pressure of an air/water gas mixture, containing O_2, N_2, and H_2O, is equal to the sum of the partial pressures of each gas:

$$p_m = p_{N2} + p_{o2} + p_{H2o} + \ldots$$

Therefore, the partial pressure of water vapor in air is directly related to measurement of humidity. This vapor pressure varies from 23.2×10^{-6} in. of mercury at the -108 °F frost point of "bone dry" arctic or industrial dry air, to 23.5 in. of mercury at the 200 °F dew point of saturated hot air in a product dryer. This is a change of a million-to-one over the span of interest in industrial humidity measurement.

The ideal humidity instrument would be a linear, wide-range pressure gauge, specific to water vapor and employing a primary or fundamental measuring method. Such an instrument, although physically possible, would be cumbersome. Most humidity measurements are made by some secondary instrument which is responsive to humidity-related phenomena.

COMMON HUMIDITY PARAMETERS

Humidity parameters most often encountered in scientific and industrial applications are shown in Table 2-2. In addition to these common parameters, dozens of other formats exist for use in narrow application or specific technologies. However, most of these are variations of the basic parameters given in the Table.

The psychrometric chart, Fig. 2-14, provides a quick means of converting from one humidity format to another since dew point, relative humidity, ambient temperature and wet bulb temperature are conveniently related to each other on a single sheet of paper. The psychrometric chart has long been the basic tool of air conditioning engineers, and Fig. 2-14 shows temperatures most often encountered in comfort control or product conditioning applications. Psychrometric charts are available for higher temperatures and humidities, and are quite useful in dryer and condensation system design. Charts are also available for lower temperatures but tend to be less useful since wet bulb measurements are difficult to make with any accuracy at temperatures below 20 °F.

Each of the four basic humidity parameters can be measured with a variety of instruments, but six standard sensors are commonly used.

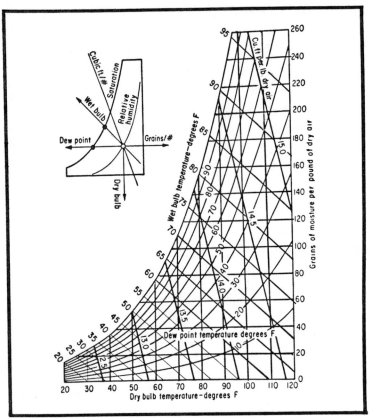

Fig. 2-14. The psychrometric chart relates all of the basic humidity parameters. Directions for reading the chart are illustrated by the inset.

Wet bulb temperature is the minimum temperature reached by a wetted thermometer in an air-stream. The humidity measurement is derived by comparing the wet bulb temperature to the ambient, or dry bulb temperature, on a psychrometic chart. Wet bulb temperatures are measured with a mechanical or electronic *psychrometer*.

Relative humidity (percent RH) is the ratio of actual water vapor pressure in the air to the water vapor pressure in saturated air at the same temperature. Percent RH can be determined from wet bulb/dry bulb or dew point/dry bulb data, but it is usually obtained directly from either a Dunmore or Poper *electric relative humidity sensor*.

Table 2-2. Humidity Measurement Methods

Parameter	Description	Units	Typical Applications
Wet bulb temperature	Minimum temperature reached by a wetted thermometer in an airstream	°F or °C	High temperature dryers, air conditioning, meteorology, test chambers
Percent relative humidity	The ratio of the actual vapor pressure to the saturation vapor pressure, with respect to water, at the prevailing dry bulb temperature	0-100%	Monitoring conditioning rooms, test chambers, pharmaceutical and food packaging
Dew/frost point	Dew point is the temperature to which the air must be cooled to achieve saturation. If the temperature is below 32 °F, it is called the frost point	°F or °C	Heat treating, annealing atmospheres, dryer control, instrument air monitoring, meteorological/environmental measurements
Volume or mass ratio	Parts per million (ppm) by volume is the ratio of the partial pressure of the water vapor to the partial pressure of the dry carrier gas. PPM by weight is identical to ppm by volume, but the ratio changes according to the molecular weight of the carrier gas.	ppm_v, ppm_w	Used primarily to insure dryness of industrial process gases such as air, nitrogen, oxygen, methane, hydrogen, etc.

Dew/frost point measurements are often obtained with the *saturated-salt* (lithium-chloride) *dew point sensor*. The dew point is the temperature at which water vapor begins to condense. If the dew point occurs below 32 °F, it is called the frost point. In applications requiring very high accuracy over a wide range, or measurements of dew points at very low humidities, a *condensation-type hygrometer* is employed.

Parts per million (ppm) measurements refer to the ratio, by volume or weight, of the actual moisture content in a carrier gas. In very dry industrial gases, ppm measurements are commonly made with an *electrolytic hygrometer* employing a phosphorous pentoxide (P_2O_5) sensor.

OPERATING RANGE AND ACCURACY

In selecting a humidity sensor for a particular application, the fundamental question is: Does the sensor operate over the span of humidity and ambient temperature which is of interest? This question usually precludes the selection of the humidity parameter to be measured since no single humidity sensor covers the entire span. Almost any of the standard humidity sensors can be used in the mid-regions of humidity and temperature, Fig. 2-15. Here the choice of sensor is dictated by the degree of accuracy or response required, the types of contaminants anticipated, and/or the cost of the system.

When temperatures extend above 140°F, or below 10° F, most relative humidity sensors offer considerably reduced performance, and other sensors are generally preferred. Likewise, as the relative humidity drops below 10 percent, neither wet bulb, electrical RH, or saturated salt dew point sensors can provide a useful measurement; either electrolytic or condensation-type dew point hygrometers provide the better choice.

Errors associated with standard humidity sensors are shown in Fig. 2-16. The figure illustrates uncertainties in mixing ratio measurements for each type of sensor over a range of dew points and mixing ratios. Mixing ratio expresses humidity as mass of water vapor per unit mass of dry air (g/kg).

At high mixing ratios, the gravimetric train is the most accurate of the sensors and its uncertainty is constant. For this reason, NBS uses the gravimetric train to calibrate other humidity instruments. At lower mixing ratios, gravimetric train errors increase at a rate which is inversely proportional to the mixing ratio.

Dew point and RH devices are characterized by constant errors in dew point temperature and percent RH measurements, respectively, across the range of mixing ratios. When expressed in terms of mixing ratio uncertainties, these constant errors show up as sloping lines.

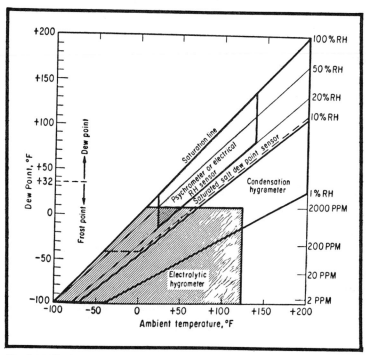

Fig. 2-15. Useful operating ranges of various humidity sensors. Psychrometers, percent RH, and saturated salt dew point sensors operate in the mid-ranges of temperature and humidity. Condensation hygrometers are useful over a large temperature span and to humidities as low as one percent. Electrolytic hygrometers operate at low dew points over the entire range of humidity.

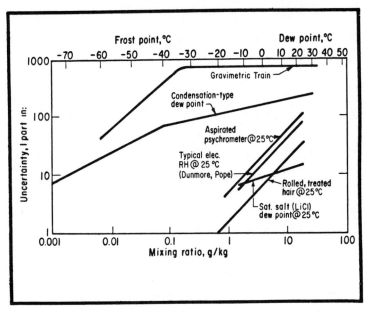

Fig. 2-16. Uncertainties in typical humidity sensors, under laboratory conditions, are illustrated here. The gravimetric train, an extremely accurate sensor which weighs actual moisture content in known air volumes, is used by NBS to calibrate other sensors. In industrial installations, few devices attain the accuracies indicated.

Figure 2-16 reflects errors inherent in the sensors themselves when properly applied in a clean environment. In essence, this error analysis represents the best performance the sensor can offer, and does not account for errors caused by contamination of the sensor, improper sampling procedure, or poor readout technique. Few sensors approach this degree of performance in actual applications, and typical performance is at least an order of magnitude less than depicted. Although comparative error analyses cannot reflect the vagaries of instrument application, the chart provides a useful means of examining relative accuracies of humidity sensors.

Every humidity sensor is subject to two major sources of error: sampling problems and contamination of the sensor. Sampling errors stem mainly from failure to locate or install the sensor such that it sees a representative sample of the gas or atmosphere being measured

154

Humidity sensors measure only the water vapor in the immediate layer of gas contacting the surface of the sensor. If the gas contacting the sensor surface has been heated by some unaccounted-for means, a percent RH sensor will reflect a lower percent RH than the true one. If a dew point sensor is mounted in a dry air line which has leaks, it can measure a dew point higher than that of the dry air. In applying any humidity sensor it is most important that the user explicity follow the manufacturer's installation instructions.

All humidity sensors are subject to errors due to contamination. Effects of comtamination vary widely with the type of sensor and the nature of the contaminant. The effects of a specific contaminant on a particular sensor are discussed as each sensor-type is reviewed in the following paragraphs.

Wet Bulb/Dry Bulb Measurements

Psychrometry has long been a popular method for monitoring humidity, primarily due to its simplicity and inherent low cost. A typical industrial psychrometer consists of a pair of matched electrical thermometers, one of which is maintained in a wetted condition.

Water evaporation cools the wetted thermometer, resulting in a measurable difference between it and the ambient, or dry bulb measurement. When the wet bulb reaches its maximum temperature depression, the humidity is determined by comparing the wet bulb/dry bulb temperatures on a psychrometric chart. In a properly designed psychrometer, both sensors are aspirated at an airstream rate between 4 and 10 m/s for proper cooling of the wet bulb, and both are thermally shielded to minimize errors from radiation.

A properly designed and utilized psychrometer, such as the Assman laboratory type, is capable of providing accurate data. However, very few industrial psychrometers meet these criteria and are therefore limited to applications where low cost and moderate accuracy are the underlying requirements. The psychrometer does have certain inherent advantages:

- The psychrometer is capable of highest accuracy near 100 percent RH. From an accuracy standpoint, it is superior to all other humidity sensors near saturation. Since the dry bulb and wet bulb sensors can be connected differentially, this allows the wet bulb depression (which approaches zero as the relative humidity approaches 100 percent) to be measured with a minimum of error.
- Although large errors can occur if the wet bulb becomes contaminated or improperly fitted, the simplicity of the device affords easy repair at minimum cost.
- The psychrometer can be used at ambient temperatures above 212 °F, and the wet bulb measurement is usable up to 212 °F.

Major shortcomings of the psychrometer are:

- As relative humidity drops to values below about 20 percent RH, the problem of cooling the wet bulb to its full depression becomes difficult. The result is seriously impaired accuracy below 20 percent RH, and few psychrometers work at all below 10 percent RH.
- Wet bulb measurements at temperatures below 32 °F are difficult to obtain with any high degree of confidence. Automatic water feeds are not feasible due to freezeup.
- Because a wet bulb psychrometer is a source of moisture, it can only be used in environments where added water vapor from the psychrometer exhaust is not a significant component of the total volume. Generally speaking, psychrometers cannot be used in small, closed volume.

PERCENT RELATIVE HUMIDITY

Percent relative humidity is certainly the most well-known, and perhaps the most widely used (and abused) method for expressing the water vapor content of the air. Percent relative humidity is defined as the ratio of the prevailing water vapor pressure, e_a, to the water vapor pressure if the air were saturated, e_s, multiplied by 100:

$$\%RH = e_a/e_s \times 100$$

The term "percent relative humidity" appears to have derived from the invention of the hair hygrometer in the 17th century. The hair hygrometer operates on the principle that many organic filaments, such as hair, goldbeater's skin and even nylon, change length as a nearly linear function of the ratio of prevailing water vapor pressure to the saturation vapor pressure. It appears that the term relative humidity was derived to accommodate the behavior of the hair, rather than to provide a desirable form for presenting humidity!

Basically, percent relative humidity is an indicator of the water vapor saturation deficit of the gas mixture, rather than an indicator of sorption, desorption, comfort or evaporation. A measurement of RH without a corresponding measurement of dry bulb temperature is not of particular value, since the water vapor content cannot be determined from percent RH alone.

Percent RH is often successfully applied where the effects of water vapor on organic materials are of interest. The sogginess and, hence, flavor of cornflakes, or the stability of grain, tobacco, and paper, relate well to a measurement of relative humidity.

SENSORS TO MEASURE PERCENT RH

Over the years devices other than the simple hair hygrometer have evolved which permit a direct measurement of percent relative humidity. These devices are, for the most part, electrochemical sensors which offer a degree of ruggedness, compactness and remote electronic readout ability not afforded by hair devices.

The two most widely used electronic percent RH sensors are the Dunmore element and the Pope cell. The Dunmore sensor employs a bifilar-wound inert wire grid on an insulative substrate which is coated with a lithium chloride solution of a controlled concentration. The hygroscopic nature of this salt causes it to take on water vapor from the surrounding atmosphere, and the AC resistance of the sensor is then an indication of the prevailing percent RH. Dunmore cells are excellent RH sensors, but—because of the characteristics of lithium chloride—are usually designed to cover a narrow range of interest. For example, a single sensor might cover from 40 to 60 percent RH, Fig. 2-17A, and the sensor output is usable only in that range.

Wide-range Dunmore sensors can be made with a cluster of narrow range sensors in a common housing, mated with an electrical matching network. This arrangement, however, usually results in a fairly bulky sensor.

The Pope cell employs a similar bifilar conductive grid on an insulative substrate; but in this sensor the substrate is made from polystyrene, which has been treated in a prescribed fashion with sulfuric acid, resulting in sulfonation of the longer chain polystryene molecules. Because the sulfate radical SO_4 is highly mobile in the presence of hydrogen ions (available from the water molecule in the vapor form) the $(SO_4)=$ ions can detach and take on $H+$ ions, thereby altering the surface resistivity of the sensor as a function of the humidity.

In both Dunmore and Pope sensors, the element is arranged in an AC-excited Wheatstone bridge, so that only alternating current flows through the grid. Direct current excitation of either the Dunmore or Pope elements polarizes the sensor, eventually causing loss of calibration.

The Pope sensor has at least one significant advantage over the Dunmore sensor in that it is a wide-range sensor, typically covering 15 percent RH to 99 percent RH in a single element, Fig. 2-17B. However, considerable attention must be given to readout circuitry for the Pope sensor, since the

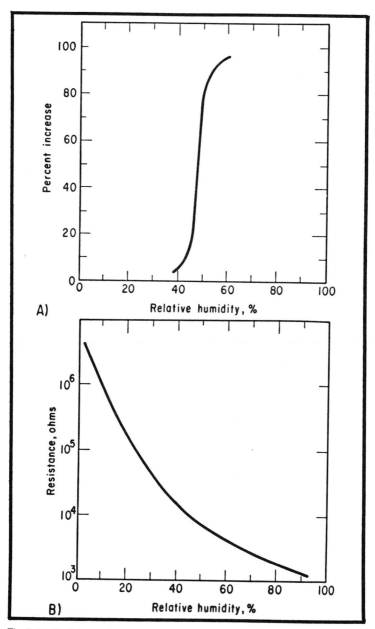

Fig. 2-17. Resistance characteristics of typical Dunmore and Pope sensors. A) Dunmore sensors are limited to a narrow range of humidity. This sensor operates between 40 and 60 percent RH. B) Pope sensors operate over a wide humidity range, but output impedance of the sensor varies from 1,000 ohms at 100 percent RH to several megohms at 10 percent RH which complicates readout circuitry.

resistance varies in a nonlinear fashion from 1,000 ohms at 100 percent RH to several megohms at 10 percent RH.

Both the Dunmore and Pope sensors are sensitive to contamination. The Dunmore sensor, if allowed to become saturated with water, tends to suffer from washout of the lithium chloride solution, resulting in loss of calibration. The Pope cell is less susceptible to wetting since the ion carriers are locked into the polystryrene substrate. Both sensor are subject to errors if contaminated with soluble salts, halogens or any material which can alter the surface resistivity of the element.

Both sensors are relatively rapid response devices—responding to changes in RH within seconds. Dunmore and Pope sensors are sensitive to changes in temperature, but can be corrected accordingly by compensating the Wheatstone bridge. Typical physical embodiments of Dunmore and Pope cells are shown in Fig. 2-18.

RH MEASUREMENT GUIDELINES

As a generalization, percent RH is the parameter to measure:

- When comfort or human performance-related measurements are needed

- When the effects of humidity on the storability, usage, consistency, etc., of dry products (paper, food, pharmaceuticals, tobacco, textiles) are of interest

- When measurements of "water activity" determined through equilibration RH are of interest

- When measurements are required in small, closed places where sensor thermal dissipation (usually a problem in dew point sensors) or injection of water vapor (as from a psychrometer) cannot be tolerated

- When fast response to a changing condition is needed.

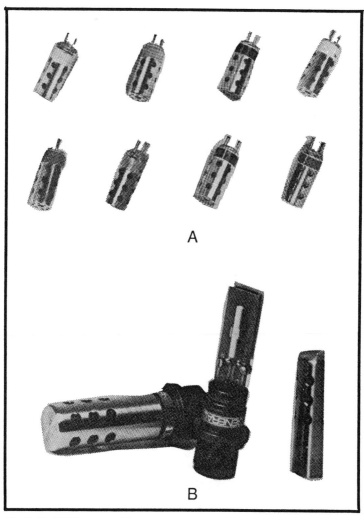

Fig. 2-18. A) This group of Dunmore sensors covers a range of 5 to 99 percent RH. B) Single Pope sensor has a range from 15 to 99 percent RH.

Percent relative humidity, as a measurand, should be avoided:

- When a precise determination of the actual water vapor content of the gas is needed (use dew point)
- When measurements at or near 100 percent RH, which usually cause condensation on the sensor, are anticipated (use wet bulb/dry bulb, or dew point)

- When measurements of RH below about 15 percent are consistently desired—percent RH sensors give consistently poor performance at the low humidities (use dew point or ppm).
- When high levels of contamination are likely to be present, ruining the sensor (use a sensor which can be cleaned and recalibrated)
- When wide changes in ambient temperatures are anticipated—relative humidity sensors, although temperature-compensated, tend to exhibit increased errors when applied over wide temperature spans.

In the second part of this article, characteristics and operational guidelines for dew point sensors and electrolytic hygrometers will be discussed, along with a look at some newer techniques.

REFERENCES

2.7 Ruskin, R. E., ed., *Principles and Methods of Measuring Humidity in Gases*, Vol. 1 of *Humidity and Moisture*, Reinhold Publishing Corp., New York, 1965.

2.8 Wexler, A., "Electric Hygrometers," NBS Circular 586, NBS, Washington, DC, 1957.

2.9 List, R.J., "Smithsonian Meteorological Tables," 6th rev. ed., Publication #4014, Smithsonian Institution, Washington, DC.

2.10 Wexler, A., *Meteoroloical Monograph*, Vol. 11, No. 33, American Meteorological Society, October 1970.

PART II: HYGROMETRY

Dew point hygrometry is widely used in scientific and industrial applications when precise measurement of water vapor pressure is needed. Dew point, the temperature at which water condensate begins to form on a surface, can be accurately measured from $-100\ °F$ to $+212\ °F$ across the entire range of humidity. Humidity measurement in very dry gases—those with dew points below $-50°F$—is sometimes difficult to achieve with dew point sensors and often calls for the use of electrolytic hygrometers, which measure the amount of water vapor in parts-per-million (ppm) concentrations.

Only two types of instruments have received wide acceptance in dew point measurement: the saturated salt dew point sensor and the condensation-type hygrometer. Many other instruments are used in specialized applications including pressure ratio devices, dewcups and fog chambers, but these are considered manual instruments, not suitable for industrial or process use.

SATURATED SALT DEW POINT SENSORS

The saturated salt (lithium chloride) dew point sensor is probably the most widely used dew point sensor due to its inherent simplicity, ruggedness and low cost. Both the U.S. and Canadian weather bureaus use this type of sensor for most official ground-based humidity measurements.

The principle of the sensor is based on the relationship that the vapor pressure of water is reduced in the presence of salt. When water vapor in the air condenses on a soluble salt, it forms a saturated layer on the surface of the salt. This saturated layer has a lower vapor pressure than water vapor in the surrounding air. If the salt is heated, its vapor pressure increases to a point where it matches the water vapor pressure in the surrounding air and the evaporation/condensation process reaches equilibrium. The temperature at which equilibrium is reached is directly related to the dew point.

A saturated salt sensor is constructed with an absorbent fabric bobbin covered with a bifilar winding of inert electrodes and coated with a dilute solution of lithium chloride, Fig. 2-19. Lithium chloride (LiC1) is often used as the saturating salt

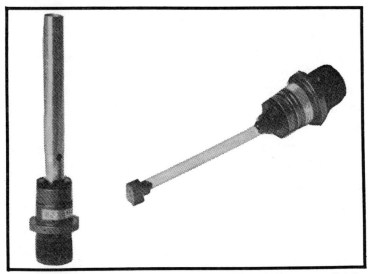

Fig. 2-19. Saturated salt dew point sensor shown with and without cover. Fabric bobbin is wound with electrodes and coated with lithium chloride (LiCl) solution. Water vapor combines with LiCl to form a saturated salt layer; electrode heats bobbin until water vapor pressure in air and in saturated layer are equal. Dew point is calculated from bobbin temperature at equilibrium.

because its hydroscopic nature permits application in relative humidities between 11 and 100 percent.

An alternating current is passed through the winding and salt solution, causing resistive heating. As the bobbin heats, water evaporates into the surrounding air from the diluted LiC1 solution; the rate of evaporation being determined by the vapor pressure of water in the surrounding air. When the bobbin begins to dry out, due to evaporation of water, resistance of the salt solution increases. With less current through the winding, because of increasesd resistance, the bobbin cools and water begins to condense, forming a saturated solution on the bobbin surface. Eventually, equilibrium is reached and the bobbin neither takes on nor loses any water.

Properly used, a saturated salt sensor is accurate to ±2° F between dew point temperatures of 10 to 100 °F. Outside these limits small errors may occur, due to multiple hydration characteristics of lithium chloride which produce ambiguous results at 106, 9 and -29 °F dew points. Maximum errors at these ambiguity points are 2.5, 3 and 6 °F, respectively, but actual errors encountered in typical applications are usually less.

SALT SENSOR APPLICATIONS

The saturated salt sensor has certain advantages over other electrical humidity sensors, such as percent RH instruments. Because the sensor operates as a current carrier saturated with L1 and C1 ions, addition of contaminating ions has little effect on its behavior compared to a typical percent RH sensor, which operates "starved" of ions and is easily contaminated. A properly designed saturated salt sensor is not easily contaminated since, from an ionic standpoint, it can be considered precontaminated.

If a saturated salt sensor does become contaminated, it can be washed with an ordinary sudsy ammonia solution, rinsed and recharged with lithium chloride. It is seldom necessary to discard a saturated salt sensor if proper maintenance procedures are observed.

Limitations of saturated salt sensors include a relatively slow response time and a lower limit to the measurement range, imposed by the nature of lithium chloride.

The sensor cannot be used to measure dew points when the vapor pressure of water is below the saturation vapor pressure of lithium chloride, which occurs at about 11 percent RH. In certain gases, ambient temperatures can be reduced, increasing the RH to above 11 percent; but the extra effort needed to cool the gas usually warrants selection of a different type of sensor. Fortunately, a large number of scientific and industrial measurements fall above this limitation and are readily handled by the sensor.

Slow response of the sensor is due to the bobbin control process, which is dependent on the thermal mass of the bobbin, amount of electrical current and flow rate of the surrounding air. Response time of the heating/cooling and

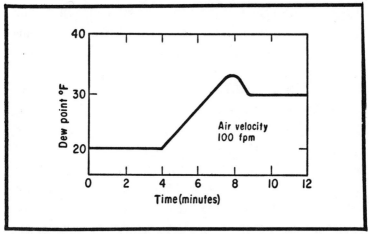

Fig. 2-20. Response characteristics of saturated salt sensor. Initially, water condenses on bobbin, forming saturated layer. Due to increased conductivity, bobbin heats, and water begins to evaporate until, after 7 min, the bobbin cools down and reaches equilibrium at 30 °F.

evaporation/condensation process is shown in Fig. 2-20. Although the response of a saturated salt sensor is slow, it is often many orders of magnitude faster than some industrial processes. The sensor is widely used in monitoring dry air systems, chambers, curing ovens, dryers, etc., where the mass of the system changes slowly and slow response in the sensor is a desirable characteristic.

Generally speaking, saturated salt sensors are a correct choice when a low cost, rugged, slow responding and moderately accurate sensor is needed. For applications requiring greater accuracies, or for humidities lower than 11 percent RH, condensation-type or electrolytic hygrometers should be considered.

CONDENSATION-TYPE HYGROMETERS

The condensation-type dew point hygrometer is one of the most accurate, reliable and wide range sensors available for humidity measurements. These features are achieved, however, through increased complexity and cost. In the condensation-type hygrometer, a surface is cooled (either thermo-electrically, mechanically or chemically) until dew or frost begins to condense out. The condensate surface is maintained electronically in vapor pressure equilibrium with the surrounding gas while surface condensation is detected by optical, electrical, or nuclear techniques, Fig. 2-21. The surface temperature is then the dew point temperature, by definition.

The largest source of error in a condensation hygrometer stems from the difficulty in measuring condensate surface temperature accurately. Typical industrial versions of the instrument are accurate to ±1 °F over very wide temperature spans. With proper attention to the condensate surface temperature measuring system, errors can be reduced to less than ±0.5 °F. Condensation-type hygrometers can be made surprisingly compact using solid-state optics and thermoelectric cooling, Fig. 2-22. Condensate detectors and control loops which maintain surface temperatures within ±0.05 °F of the true dew point are available.

Wide span and minimal errors are two of this instrument's main features. A properly designed condensation hygrometer

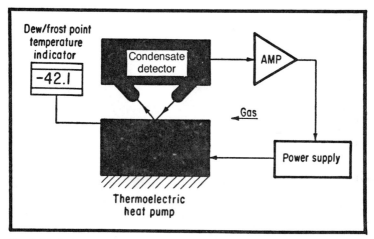

Fig. 2-21. Dew point is detected in a condensation hygrometer by cooling a surface until water begins to condense. Condensation is detected optically or electronically, and the signal is fed into a control circuit which maintains the surface temperature at the precise dew point.

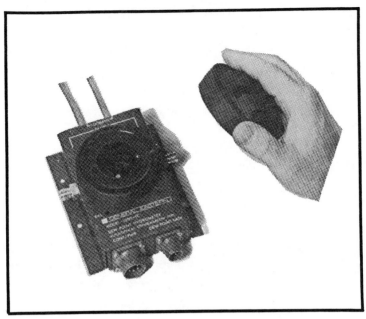

Fig. 2-22. Typical condensation hygrometer with optical detector.

can measure dew points from 212 °F down to frost points of −100 °F. Typical condensation hygrometers can cool to 120° F below the ambient tempterature, establishing lower limits of the instrument to dew points corresponding to 1-2 percent RH. Accuracies for measurements above −40 °F can be ±1 °F or better, deteriorating to ±2 °F at the lower temperatures.

Response time of a condensation dew point hygrometer is usually specified in terms of its cooling/heating rate, typically 2 °F/s, making it somewhat faster than a saturated salt dew point sensor and nearly as fast as most electrical percent RH sensors. Perhaps the most significant feature of the condensation hygrometer is its fundamental measuring technique, which essentially renders the instrument self-calibrating. For calibration, it is only necessary to manually override the surface cooling control loop, causing the surface to heat, and witness that the instrument re-cools to the same dew point when the loop is closed: assuming that the surface temperature measuring system is calibrated, this is a reasonably valid check on the instrument's performance.

Another feature of the condensation hygrometer is its inert construction, which renders it virtually indestructible. Although condensation hygrometers can become contaminated, they can easily be washed and returned to service with no impairment in performance.

APPLICATIONS OF CONDENSATION HYGROMETERS

Perhaps the major disadvantage of the condensation hygrometer is its complexity and consequent cost, when compared to other humidity sensors.

A second disadvantage is that the instrument is a source of heat and cannot always be used in small, closed containers. It is customary to draw the sample to be measured through the sensor, using a pumped sampling system, which adds to the cost of the measurement.

As with all hygrometers, the condensation type is also subject to contamination. Materials other than water on the condensing surface cause the dew point reading to be altered. Problems are experienced when soluble salts (limestone, sea salt, etc.) or condensible vapors (such as hydrocarbons) are present, since the instrument tends to sense and control on the highest vapor component, which may not be water vapor. Certain types of condensation hygrometers which utilize condensation detection schemes specific to water can overcome this disadvantage.

A condensation-type dew point hygrometer is best suited whenever maximum accuracy of water vapor content is needed over wide ranges of dew point. Also, in applications where there is a chance of routinely contaminating the sensor with oils, corrosive gases or salts, the condensation-type hygrometer is usually the correct choice because of its serviceability.

Generally speaking, condensation-type hygrometers work well on clean gases all the way down to $-100\ °F$ frost

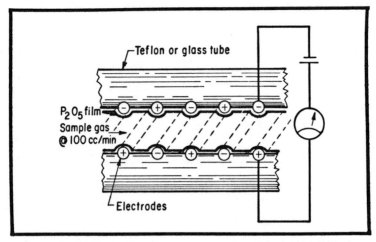

Fig. 2-23. Electrolytic hygrometer dissociates water, absorbed by P_2O_5, into hydrogen and oxygen by electrolysis. Since two electrons are required to electrolyze a molecule of water, the amount of current used by the hygrometer relates to ppm of water vapor.

points. However, below about -40 °F, response of the instrument is considerably slowed, and susceptibility to errors due to contamination is greatly increased. Also, the task of cooling the surface from high ambients to a low temperature usually requires ancillary coolant, resulting in additional instrument complexities. Below frost points of -50 °F, other instruments such as the electrolytic hygrometer are often the better choice.

ELECTROLYTIC HYGROMETERS

Since the electrolytic hygrometer is specific to water and impervious to most contaminants, it is often used for water vapor measurements in dry gases. An electrolytic hygrometer electrolyzes water vapor into its components, hydrogen and oxygen: the amount of electrical current required to dissociate water vapor at a particular temperature and flow rate is proportional to the number of water molecules present in the sample.

A typical electrolytic hygrometer utilizes a cell coated with a thin film of phosphorous pentioxide (P_2O_5), which absorbs water from the sample gas, Fig. 2-23. The cell has a bifilar winding of inert electrodes on a fluorinated hydrocarbon capillary—direct current applied to the electrodes dissociates the water, absorbed by the P_2O_5, into hydrogen and oxygen. Two electrons are required to electrolyze each water molecule, and the current in the cell represents the number of molecules dissociated. A further calculation, based on flow rate, temperature and current, yields the parts-per-million concentration of water vapor.

In order to obtain accurate data, the flow rate of the sample gas through the cell must be known and constant. Since the ppm calculation is partially based on flow, an error in the flow rate causes a direct error in measurement.

A typical sampling system to insure constant flow, Fig. 2-24, maintains constant pressure within the cell. Sample gas enters the inlet, passes through a stainless steel filter and enters a stainless steel manifold block. (It is most important that all components prior to the sensor be made of an inert material, such as stainless steel, to minimize contamination.)

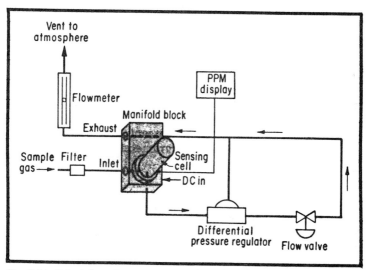

Fig. 2-24. Calculation of water vapor content in an electrolytic hygrometer is dependent on precise control of the flow rate. This arrangement controls ample pressure across the cell, insuring correct flow regardless of input pressure fluctuations.

171

After passing through the sensor, the sample gas pressure is controlled by a differential pressure regulator which compares pressure of the gas leaving the sensor to gas venting to atmosphere through a preset valve and flowmeter. Thus, constant flow is maintained in spite of nominal pressure fluctuations at the inlet port.

Typical minimum sample pressure required for this type of control is 10 psig ($1.7 kg/cm^2$) to 100 psig ($8 kg/cm^2$). If sample pressure is more than 100 psig, an external upstream regulator is employed. A sintered stainless steel filter is also employed upstream to prevent contaminants from reaching the cell, if necessary.

A typical electrolytic hygrometer can cover a span from 0 to 2,000 ppm, with an accuracy of ± 5 percent of the reading, more than adequate for most industrial applications. The sensor is suitable for most inert elemental gases and organic and inorganic gas compounds that do not react with P_2O_5.

GASES TO AVOID

The electrolytic hygrometer cannot be used for humidity measurement on gases which are corrosive (chlorine, etc.) or which readily combine with P_2O_5 to form water (alcohols). Also, gases such as certain acids, amines and ammonia react with P_2O_5 and must be avoided. Unsaturated hydrocarbons (alkynes, alkadienes and alkenes higher than propylene) polymerize to a liquid or solid phase, clogging the cell.

Unlike condensation hygrometers, P_2O_5 cells are not inert and cannot easily be cleaned if clogged or short-circuited. Short-circuited cells must be discarded. Clogged cells can occasionally be cleaned and reactivated with P_2O_5.

Electrolytic hygrometers cannot be exposed to high water vapor levels for any long period of time, as this results in high usage rates for the P_2O_5 and high cell currents.

DRY GAS MEASUREMENT PROBLEMS

Measurements of water vapor in very dry gases—dew points below $-50\,°F$ or 80 ppm water vapor by volume—pose relatively difficult instrumentation problems. Only two types of commercial instruments are available for such measurements: condensation and electrolytic hygrometers.

A fundamental problem in dry gas measurements is that sampling hardware must not entrain water, which would cause higher than actual measurements. A second problem is that the small amount of water vapor available is often of the same magnitude as other gas constituents with similar vapor pressure characteristics.

In application of a condensation hygrometer at $-50\ °F$ dew point, for example, nearly 100 cc of gas must be passed over the cooled surface to condense out the few (typically, 10) micrograms of condensate needed to identify the dew point. In this same volume of gas, there is likely to be an even greater mass of oil or other hydrocarbon vapor condensing out onto the surface, masking the water vapor. If the gas is clean, this is not a problem. But if not, as is often the case in industrial applications, the condensation hygrometer becomes unusuable at low dew points.

Measuring water vapor content in very dry gases also requires a considerable time for the sample lines to come into equilibrium with the gas. Neither the electrolytic hygrometer nor the condensation hygrometer are rapidly responding devices at low dew points, and an accurate measurement of moisture content in the 0 to 50 ppm range can take *several hours*. The major factor contributing to such slow response is the time required for the instrument and sample lines to "dry down" so that they are in water vapor equilibrium with the gas sample. Use of short sampling lines fabricated from nonhydroscopic materials (preferably, stainless steel) operating at high flushing velocities are mandatory requirements of sampling systems used at low water vapor levels.

Lastly, measurement of dew point made at other than atmospheric pressure must be corrected, or the total pressure must be stated along with the water vapor value. Dew point is a function of pressure (increasing the total pressure raises the water vapor pressure and, hence, the dew point), and it has become conventional to assume dew point readings are referenced to atmospheric pressure when no other pressure is stated.

OTHER HUMIDITY SENSORS

Certain newer techniques have been developed and introduced during the last decade. Two of these new methods are the aluminum oxide and Brady Array sensors.

An aluminum oxide sensor is a capacitor, formed by anodizing a strip of aluminum with a porous oxide layer and then coating the oxide with a thin film of gold. The aluminum base and the gold layer become the capacitor's electrodes. Water vapor penetrates the gold layer and is absorbed by the porous oxidation layer. The number of water molecules absorbed determines the conductivity and electrical impedance of the capacitor which is, in turn, a measure of water vapor pressure.

A Brady Array sensor is a solid-state semiconductor device consisting of a precise array of crystals and interstitial spaces which absorb water molecules. The sensor is mounted in a slotted TO-5 transistor enclosure. A 5-V, 1,000 Hz input signal is modified by the array, producing a 0-5 Vdc output according to the amount of water absorbed by the array.

Both of these sensors were recently tested by NBS. The aluminum oxide sensor was found to have an inherent uncertainty of 4 °C, and hysteresis of 4 °C at dew points between −4 and +18 °C. Over a five-month period of use, calibration can change by as much as 12 °C.

The Brady Array was found to be temperature-dependent, requiring calibration at the precise temperature at which it would be used. Sensitivity decreases at lower temperatures, and at temperatures below −20 °C, the device is unusable. Short- and long-term repeatability was poor, with errors up to 10 percent, and hysteresis was found to be 9 percent and higher. Aluminum oxide and Brady Array sensors may be useful in some applications where precise measurements are not needed and large hysteresis effects can be tolerated.

BIBLIOGRAPHY

Hasegawa, S., *et al*, "A Laboratory Study of some Performance Characteristics of an Aluminum Oxide Humidity Sensor," NBS Technical Note 824, National Bureau of Standards, Washington, DC, 1974.

Hasegawa, S., *et al*, "Performance Characteristics of a 'Bulk Effect' Humidity Sensor," NBSR #74-477, National Bureau of Standards, Washington, DC, 1974.

PIETER R. WIEDERHOLD is President of General Eastern Corp., Watertown, MA. Article is based on a paper presented at the 21st ISA Analysis Instrumentation Symposium, 1975, King of Prussia, PA.

Selecting the Right Flowmeter

D. J. LOMAS

PART I: THE SIX FAVORITES

Flow measurement is a complex subject: not only are many types of flowmeters available, but selecting one for a particular application depends on many factors. A great deal of confusion exists as to where and when each type of meter should be used.

This article concentrates on application considerations for six of the most popular types of flowmeters. Their advantages, features and—perhaps more important—their limitations will be presented, along with comparative data on major selection criteria such as accuracy, cost, installation requirements, flow ranges, pressure drop, etc. The six flow metering methods considered are:

- differential pressure
- positive displacement
- electromagnetic
- turbine meters
- rotameters
- vortex meters.

In four of the categories, "point velocity" devices are also available such as pitot tubes, insertion electromagnetic meters, insertion turbine meters and insertion vortex meters. Such systems do not measure total flow, but infer it from a velocity measurement taken at one point in the pipe velocity profile. They assume a standard distribution pattern which does not change with flow rate or variations in operating conditions, severely limiting their use in industrial applications. Consequently, point velocity devices are not considered in this article.

DIFFERENTIAL PRESSURE METERS

Probably the most widely used method of industrial flow measurement, differential pressure (ΔP) meters are available in a variety of types—orifice plate, venturi tubes, Dall tubes, etc.—but all operate on the principle that fluid accelerated through a restriction has its kinetic energy momentarily increased at the expense of its pressure energy, Fig. 2-25. The differential pressure h caused by a velocity increase varies according to the square of flow rate Q:

$$Q = K \sqrt{h/d}$$

where d is the density, and K is a constant.

This formula illustrates two major limitations which apply to all ΔP systems:

- Flow measurement range is severely limited by the square root relationship between the head and flow rate.
- Density must be known or measured.

In terms of rangeability, ΔP systems are limited to a range of 4:1 (at the most, 5:1). Because of the square root relationship, a 5:1 flow range means a 25:1 turndown ratio on measured head—and this measurement is transmitted by an instrument whose accuracy is specified as percent fullscale. The resulting system accuracy, particularly at reduced flow rates, can be adversely affected.

178

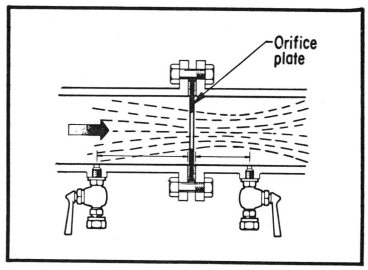

Fig. 2-25. Fluid accelerated through a restriction suffers a pressure drop; this pressure drop is directly related to flow rate by a differential pressure flowmeter.

The importance of density change is not applicable in many liquid applications; however, density changes in gaseous fluid metering can cause significant errors.

A principal advantage of ΔP meters is that they can be applied to a vast number of flow measurements involving most liquids and gases, including dirty and viscous fluids, over an extended range of pressures and temperatures. Difficult fluids in fact normally impose more severe restrictions on the secondary device used to measure the differential head than on the primary unit itself. For difficult fluids, accessories such as purging probes, sealing chambers, settling chambers and gas collectors are often required.

Orifice plates are the simplest and least expensive ΔP devices; also, their cost is relatively independent of pipe size. With most of the flowmeters, cost increases significantly with pipe size. A summary of orifice plate ΔP transmitter system characteristics, advantages and limitations is given, Table 2-3.

A principal disadvantage of the orifice plate is its high unrecoverable pressure drop (between 40 percent and 80 percent of the differential pressure generated). To overcome

Table 2-3. Orifice Meters

Advantages	Limitations
No moving components	Square root head/flow relationship
Available in a wide range of sizes and constrictions	Usable flow range limited to 4:1
Suitable for most gases and liquids	Low accuracy
Price virtually independant of pipe size	Accuracy deteriorates with wear and damage
Widely established and accepted	Accuracy affected by density and flow profile
No product lubricity problems	High unrecoverable pressure drop
Orifice need not be flow calibrated	Viscosity affects flow range
Simplicity	Maintenance is required
	Installation is time-consuming and expensive

this limitation, ΔP devices such as the Dall tube, Fig. 2-26, are available. Dall tubes use specially engineered and designed shapes so that the fluid can flow smoothly through the tube at a much higher velocity without the turbulence associated with an orifice plate.

This savings in pressure drop is made, however, at a considerable increase in cost, bulk and inflexibility. These tubes therefore are used when they result in lower capital expenditures for pumping equipment, or when savings in energy justify the initial higher cost. An indication of the dramatic savings in pressure loss which can be achieved with the low loss tubes is shown in Fig. 2-27.

TRAPPING AND PASSING

Positive displacement meters, Fig. 2-28, operate by trapping a known volume of fluid, passing it from inlet to outlet, and counting the number of fluid "packets" that pass. An output shaft drives through gearing to a local display counter; by selection of suitable gearing, readout in the required volumetric units can be obtained. A pulse generator, either optical or electromagnetic, also may be fitted for transmission to a remote control room. Because of production tolerances, individual meters must be calibrated to establish true swept volume. Fine adjustment of the meter reading, to compensate for manufacturing tolerance variations and mechanical wear, also is possible with suitable gearing.

A positive displacement (PD) meter is extremely accurate and repeatable, providing that it is adequately maintained. In fact, on high viscosity liquids, a PD meter offers unequalled accuracy and flow range. As liquid viscosity increases, the slippage and hence the error is reduced, and the meter can be operated at lower flow rates.

Because of its high performance, PD meters are widely used for flow measurement of fuel oils and other hydrocarbon products in small pipe sizes. Also, because a PD meter is

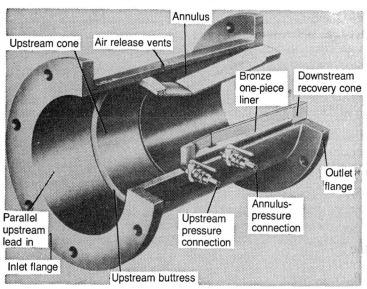

Fig. 2-26. A Dall tube can be used when an orifice plate causes too much of a pressure drop. Fluid flows smoothly through the restriction in a Dall tube and does not create turbulence. Dall tubes, however, are expensive.

181

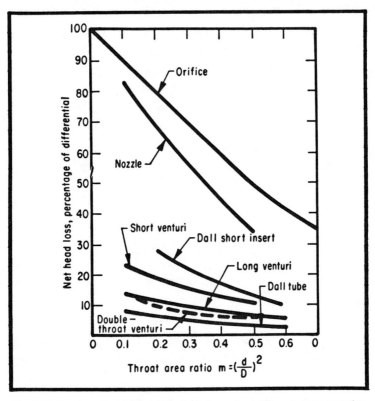

Fig. 2-27. A variety of methods are available to produce differential pressure; the pressure drops caused by each are shown here. The orifice plate causes the most drop, while the Dall tube causes the least.

self-contained and requires no electrical power, it is ideally suited for vehicular applications such-as commercial distribution of heating oil from a tank truck.

The basic limitation of a PD meter is that it has moving parts with close clearances, effectively limiting its use to clean liquids and necessitating regular maintenance on the meter. Also, the choice of suitable materials is limited, restricting the meter's corrosion resistant properties. High temperatures and pressures also can result in distortion problems unless a double casing meter is used, further increasing the cost of an already expensive meter.

For larger size PD meters the sheer physical size, bulk and weight becomes increasingly prohibitive, resulting in special requirements such as mounting pads for meter installation. These drawbacks, along with limited throughput of a given size meter, result in large space requirements in major installations such as a crude oil loading terminal. Table 2-4 details the principal advantages and limitations associated with PD flowmeters.

ELECTROMAGNETIC FLOWMETERS

Faraday established that the emf induced in a conductor moving through a magnetic field is proportional to the velocity of the conductor—and this is the principle on which the electromagnetic flowmeter is based.

An electromagnetic flowmeter, Fig. 2-29, consists of a nonmagnetic stainless steel pipe lined with an insulating material. A magnetic field is produced across the tube by exciting coils arranged around the outside. Liquid passing through the

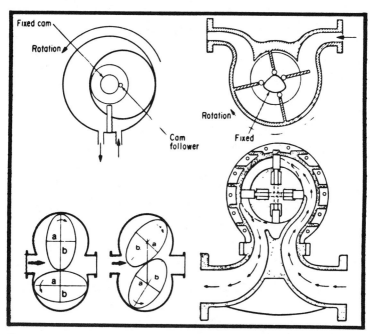

Fig. 2-28. All positive displacement flowmeters drop a quantity of fluid and pass it to the output side; flow is calculated from the number of fluid "packets" that pass in a time interval.

Table 2-4. Positive Displacement Meters

Advantages	Limitations
Good accuracy and rangeability	Close tolerance moving components subject to wear
Very good repeatability	Regular maintenance and service required
Suitable for high viscosity fluids	Not suitable for dirty, nonlubricating or abrasive liquids
Will accommodate large viscosity change	
Local readout—with option of pulse output	Expensive, particularly in large diameters
	Large and bulky for big diameters
Readout directly in volumetric units	Limited throughput for a given size
No power supply required	Expensive spares
Accuracy virtually unaffected by upstream pipe work	Difficult installation
	High head loss
High pulse resolution possible	Damaged by overspeeding
Wide flow range	

flowmeter becomes the conductor; the emf induced by the fluid in the magnetic field is proportional to the fluid velocity.

Two small electrodes opposite each other and flush with the lining detect the emf. The induced emf is not in a directly usable form since it is small in value and has a high source impedance. The signal is fed into a unit which amplifies the emf and converts it into a standard mA analog signal suitable for use with standard receiving and control units. In addition, the converter rejects unwanted signals generated in the detector head or caused from an external source and compensates for variations in supply voltage and frequency.

Because the detector head is just a lined piece of pipe with two small measuring electrodes, there is no obstruction to flow, and pressure loss through the meter is no greater than an equivalent length of straight pipe. Also, because of the obstructionless nature of the meter, slurries, pulps and liquids containing solids can be handled effectively.

Linings available include abrasion resistant rubber, polyurethane and Teflon. The standard electrode material is stainless steel; but since the electrodes are very small, they can be manufactured in more exotic materials such as tantalum or platinum iridium at relatively little increase in cost. This means that electromagnetic flowmeters can be made from materials which are chemically compatible with virtually all liquids. Consequently, the electromagnetic flowmeter can be used on many extremely corrosive or aggressive liquids which are difficult or impossible to meter with most other types of flow measurement equipment.

The principal limitation with the electromagnetic flowmeter is that the meter is not suitable for gas or nonconductive liquids. Advantages and limitations of these flowmeters are listed in Table 2-5.

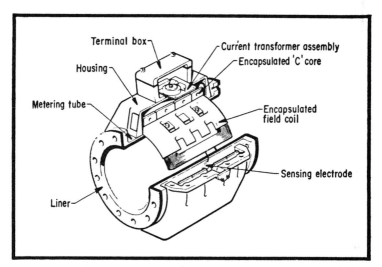

Fig. 2-29. Electromagnetic flowmeters detect emf changes in an electrically charged fluid passing through a magnetic field. The obstructionless nature of an electromagnetic flowmeter make it ideal for slurries, pulps and liquids containing solids.

Table 2-5. Electromagnetic Meters

Advantages	Limitations
No moving components	Liquid must be electrically conductive
Completely unobstructed bore	Not suitable for gas
Chemically compatible with virtually all liquids	Expensive, particularly in small sizes
Suitable for slurries, nonlubricating or abrasive liquids	Power supply required at measurement point
Unaffected by viscosity, pressure, temperature, density or conductivity	Hazardous area location is difficult
No viscosity limitation	Drift may necessitate checking
Linear analog output	Can be sensitive to asymmetric flow profile
Extensive range of sizes and flow rates	Limited on maximum temperature
Cost does not increase with size as rapidly as many flowmeters	Calibration required
Bidirectional flow standards	Possible electrode fouling in some applications
Can be re-arranged on site	

IMPINGING FLOW

A turbine meter consists basically of a bladed rotor suspended in a fluid stream with its axis of rotation perpendicular to the flow direction, Fig. 2-30. The rotor is driven by liquid impinging on the blades; the rotor's angular velocity is proportional to fluid velocity which, in turn, is proportional to the volume flow rate.

A pickup coil on the outside of the meter body detects rotor motion. The pickup coil has a magnet and magnetic field and the rotor blades are made from ferrous material; as each rotor blade passes the pickup, it cuts the magnetic field and produces a pulse. The output signal is a continuous sine wave

186

voltage pulse train with each pulse representing a small discrete volume of liquid. Associated electronic units then display total volumetric flow or flow rate, and perform preset batching, automatic temperature correction and other functions.

Although the basic theory of the turbine meter is extremely simple, the detailed design is a highly involved and complex exercise; final performance depends on a multitude of factors such as blade angle, bearing design and number of blades, as well as close tolerance machining and assembly. A correctly designed and manufactured turbine meter has high accuracy over a 10:1 flow range and excellent repeatability. In addition, a turbine meter is small and lightweight (relative to the pipe), and has high throughput for a given meter size. Installation of a turbine meter is a simple operation. Consequently, turbine meters are widely used for high accuracy royalty and custody transfer measurement of such products as crude oil or petroleum.

A turbine meter is a versatile unit: it has an extremely wide temperature and pressure range and, since it is manufac-

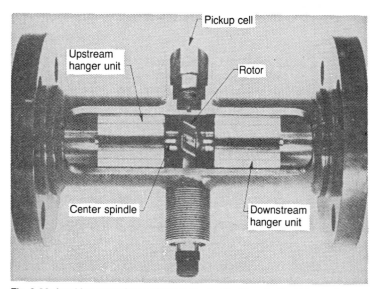

Fig. 2-30. A turbine meter has a bladed rotor that is driven by liquid impinging on the blades. A pickup coil detects rotor motion. Turbine meters are versatile, but must be used with relatively clean, low viscosity fluids.

Table 2-6. Turbine Meters

Advantages	Limitations
High accuracy	Not suitable for high viscosity
Very good repeatability	Calibration required
10:1 flow range	Can be damaged by overspeeding or gassing
Versatile and suitable for operation under severe conditions	Relatively expensive
Suitable for virtually unlimited pressure and very wide temperature range	Back-pressure requirements are high
Wide range of sizes with high maximum throughput	Moving component subject to wear
High reliability (only one moving component)	Affected by upstream flow conditions
Linear digital output	Requires secondary readout
Fast response	Filtration required
Small size and weight	
Easy installation	
Can be made hygienic	

tured in stainless steel, is compatible with a wide range of fluids. Fluids, however, must be relatively clean and not be a high viscosity product.

One potential limitation with a turbine meter is that it has a moving component—the rotor—and a bearing which is subject to wear. By using tungsten carbide for bearings, bearing life in excess of five years without maintenance can be achieved in nonlubricating liquid applications. Table 2-6 shows advantages and limitations of turbine meters.

ROTAMETERS

Variable area flowmeters—also called rotameters—utilize the same principle as differential pressure meters; i.e., the relationship between kinetic energy and pressure energy. In the ΔP system, the restriction size is fixed and the differential pressure changes according to flow rate; in the variable area meter, the area of restriction changes as the flow rate changes and the pressure differential remains constant.

A variable area flowmeter, Fig. 2-31, consists of a vertical, tapered glass tube containing a float. The float has the same diameter as the glass tube at its base. The tube tapers outward, so it is larger in diameter at the top than at the base. Fluid flows through the tube from bottom to top, carrying the float upward. As the float rises, the annular space increases until the lifting force produced by the differential pressure across the upper and lower float surfaces equalizes.

At low flow rates, the float will hover near the base of the tube; at higher flow rates, the float will rise up the tube. The equilibrium position reached by the float in the tube gives a direct linear indication of flow rate. In its simplest form, a rotameter has a glass tube with a scribed scale, and flow can be read directly. To minimize friction between the float and the

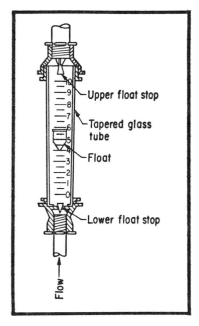

Fig. 2-31. In a variable area flowmeter (rotameter), differential pressure generated by fluid flowing into a tapered glass tube raises a float to a position that relates directly to flow rate.

Table 2-7. Variable Area Meters

Advantages	Limitations
Low cost	Not suitable for high pressures
Simplicity	
	Limited maximum flow capability
Relatively unaffected by approach pipework	
	Must be mounted vertically
Suitable for very low flow rates	
	Unit is bulky
Constant low pressure loss	
	Dirt buildup on glass can make reading difficult
10:1 flow range	
Direct linear reading of flow rate	Expense increased considerably with extras such as protection shields, panel mounting, etc.
Power supply not required	
	Transmission and integration not available as standard
	Clean fluids only

tube, the float either has inclined grooves which cause it to spin and remain in the center of the tube, or it runs on guides.

Rotameters have typical accuracy of ±2 percent over a 10:1 flow range. Temperature range is good (-46 to $+205°$ C), but the maximum operating pressure is restricted to approximately 300 psig. A significant advantage is that the meter can be used for very low flow rates, such as 5 cc/min, but its maximum flow is only 250 gpm. Other major virtues, Table 2-7 include simplicity, very low cost and a direct readout that does not require a power supply or secondary readout units.

Although approach piping configurations have little effect on performance, the meter must be mounted vertically. The meter is fairly bulky and should be protected by a strainer; if the fluid is dirty, the glass tube can become coated making it difficult to read the scale. The glass tube also has strength and sealing problems, limiting its size and pressure capabilities, and is subject to damage from thermal shocks or pressure hammering. In such circumstances, a protective shield should be used around the tube.

Using a steel tube instead of glass considerably extends the meter's scope in terms of pressure, temperature and size. With a steel tube, the float can be fitted with a rod and pointer that projects upward into suitably protected gauge glass, or the position of the float can be detected with a magnetic coupling.

Such modifications, however, detract from the meter's two principal virtues—simplicity and low cost. Rotameters are widely used in applications where only local flow indication is required for small volumes of clean fluids. Output of the meter is not suitable for transmission and the meter, therefore, is rarely used in control applications.

SHEDDING VORTEXES

Vortex meters are based on a natural phenomenon known as vortex shedding. When a fluid flows past an obstruction, Fig. 2-32, boundary layers of slow-moving viscous fluid are formed along the outer surfaces. If the obstacle is unstreamlined, i.e., a bluff body, the flow cannot follow the obstacle contours on the downstream side; the separated layers become detached and roll themselves into vortexes in the low pressure area behind the body. Vortexes are shed from alternative sides; the frequency at which they are shed is directly proportional to velocity, thus providing the basis of a flowmeter.

As a vortex sheds from one side of a bluff body, the liquid velocity on that side increases and the pressure decreases; on the opposite side, a velocity decrease and pressure increase occurs, resulting in a net pressure change across the bluff body. The entire effect is then reversed as the next vortex sheds from the opposite side. Consequently, the velocity and

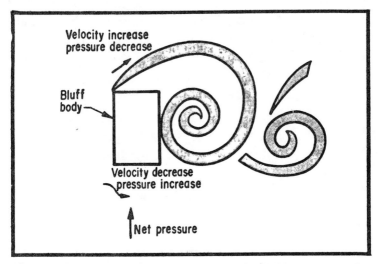

Fig. 2-32. When a vortex sheds from one side of a bluff body, the velocity increases and the pressure decreases; on the other side of the bluff body, the velocity decreases while the pressure increases. This change in pressure is sensed and related to flow rate by a vortex meter.

pressure distribution around the bluff body changes at the same frequency as the shedding frequency and various methods can be used to detect either the pressure or velocity change.

A capacitance vortex meter, Fig. 2-33, is a bluff body with a capacitance detector. Two small flats are machined on the sides of the body, interconnected by several small transfer ports. The block material is insulated with a resin, and an electrode is mounted on each side. Diaphragms are welded onto each side, and the entire assembly is filled with oil. Net pressure changes across the bluff body deflect the diaphragms, changing the capacitance between the diaphragms and the electrodes. The capacitance increases on one side and decreases on the opposite side. These changes are detected and processed to provide either a frequency or analog output signal.

A full bore vortex meter combines in one meter many advantages and features not found collectively in any other flowmeter, Table 2-8. Its principal attributes are good accuracy, long-term repeatability and good rangeability. In addition, calibration is independent of viscosity, density, pressure and temperature, and can be maintained for long periods. This is because the calibration, or meter factor, is determined only by the size and shape of the obstruction, and is a natural phenomenon. There are no moving components such as bearings or gears and, since it is a frequency system, there are no drift problems.

One remarkable feature is that the same vortex meter can be used on liquid or gas, and the calibration remains the same for either fluid—within normal accuracy tolerances. Since the vortex meter is a universal design which does not change from application to application, the spare parts inventory for a complete plant is simple.

Despite these features, the vortex meter is a low-cost unit, in terms not only of purchase cost, but also of installation

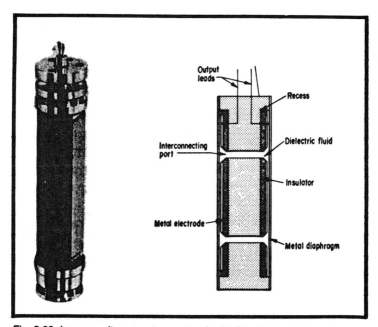

Fig. 2-33. In a capacitance vortex meter, the bluff body has diaphragms and electrodes installed on opposite sides. When a vortex sheds, the diaphragms deflect and change the capacitance. These changes are detected and output as a frequency or voltage signal.

Table 2-8. Vortex Meters

Advantages	Limitations
Low installed cost	Not suitable for dirty or abrasive fluids
Good accuracy	
Good long-term repeatability	Not suitable for viscous liquids
Generally wide flow range	Limited choice of materials of construction
Minimal maintenance	Limited size range
Calibration not required	Limited maximum pressure and temperature capability
Calibration unaffected by viscosity, density, pressure and temperature	Limited pulse resolution
Suitable for gas or liquid	Pressure drop of two velocity heads
Frequency or analog output	
Simple, interchangeable spare parts	
Analog unit can be re-ranged on site	
Simple installation	

and operating costs. Vortex meters consequently are now being consisdered as viable technical and commercial alternatives to orifice plates in many applications. Because of cost, the meter is not competitive in sizes above 6 in. or in temperatures outside a −40 °C to +120 °C range. Likewise, the vortex meter is not effective in viscous, dirty, abrasive or corrosive fluids, or where a pressure drop of two or more velocity heads is not acceptable.

In the mainstream of general applications, however, such as service metering and general process metering, the vortex meter is gaining increasing acceptance and use.

PART II: COMPARING CANDIDATES

Each of the six flowmeters described in Part I has individual advantages and disadvantages, but no one meter incorporates all the good features of the others. Consequently, all six—differential pressure, positive displacement, electromagnetic, turbine, variable area and vortex meters—continue to be used for applications in which they seem to be best suited. The question that must be answered by an engineer is: How does one determine which flow-meter is best suited for a particular application?

By using the characteristics of the fluid being measured—cleanliness, conductivity and viscosity—it is possible to quickly eliminate certain meters and narrow down the choice of meters capable of performing the task to a more manageable number. Figure 1 offers a flowchart approach, based on fluid characteristics, that can be used as a quick selection guide.

After the meters capable of performing the measurement are identified, the engineer must choose among them by comparing their respective specifications to the application operating conditions. These conditions include, but are not limited to, pressure, temperature, flow range and pressure drop. Table 2-9 can be used for a general comparison of specifications; however, in a general chart like this, specifications are subject to considerable variations because of many factors and relationships.

For example, the effects of pressure and size on an electromagnetic (EM) flowmeter are not apparent in Table 2-9. A 2-in. EM flowmeter suitable for operation at 2,000 psig is readily available. A 68-in. EM detector head suitable for operation at 20 psig is readily available. But a 68-in. detector head that can operate a 2,000 psig is a different proposition. Although the possibility of requiring a 68-in. EM flowmeter for a 2,000 psig application is virtually nil, this example illustrates the shortcomings of a generalized table.

Table 2-9. Meter Characteristics

Feature	Orifice Transmitter	Positive displacement	Electro-magnetic	Turbine meter	Vortex meter	Variable area
Accuracy/ linearity	±1% of flow ±0.5% FSD	±0.25% of flow	±1% of flow or ±0.2% FSD, whichever is the greater	±0.25% of flow	±0.5% of flow to a minimum Rd of 30,000	±2% FSD
Repeatability	--	±0.05%/0.02%	±0.2% approx.	±0.05/0.02%	±0.15%	±1%
Flow range	4:1	15:1	10:1 but can be re-ranged	10:1	Varies but typically 12:1	10:1
Minimum flow velocity	Varies	Typically 1 ft/s	0.2 ft/s	Typically 3 ft/s	Rd 10,000	Varies
Maximum flow velocity	Varies	Typically 15 ft/s	30 or 60 ft/s (limited by converter)	Typically 30/35 ft/s (limited by cavitation or bearing overspeed)	Limited by cavitation	Varies
Pressure loss	Varies but generally greater than 4 velocity heads	Typically 1-1.5 velocity heads	Zero	Typically 0.5-1.5 velocity heads	2 velocity heads	Varies (but low)

Feature	Orifice Transmitter	Positive displacement	Electro-magnetic	Turbine meter	Vortex meter	Variable area
Size range	1 in. upward	1 in./16 in.	1/8 in./72 in.	3/16 in./24 in.	2 in./6 in.	Up to 2 in. glass, 3 in. metal
Calibration	Not required	Required	Required	Required	Not required	Not required
Output	Analog (sq. rt.)	Local display or transmission option	Analog	Frequency	Analog or frequency	Local or transmission option
Secondary units	ΔP transmitter	None	Converter	Frequency electronics	None or frequency electronics	None
Service	Gas or liquid	Gas or liquid	Liquid	Gas or liquid	Gas or liquid	Gas or liquid
Maximum temperature	Above 300 °C	150 °C	120 °C	Above 300 °C	120 °C	100 °C glass, 350 °C metal
Maximum pressure	Above 1,500 psi	Varies with type	Varies with size	Above 1,500 psi	1,000 psi	150 psi glass, 1,000 psi metal

By following the flowchart in Fig. 2-34 and comparing general specifications from Table 2-9, a list of possible meters can be established. The next task is to precisely identify measurement requirements in terms of:

- accuracy
- cost
- reliability
- installation
- maintenance
- required display.

Each item on the list has various degrees of importance to the selecting engineer, depending on application requirements, and a suitable "weighting" should be assigned to each criterion.

OVEREMPHASIZING ACCURACY

The question of accuracy inevitably is one of the first factors to be considered when selecting a flowmeter. This is not surprising because a fluid generally is metered either for revenue/sale purposes or because its measurement improves process efficeincy, enables automation to be introduced, or enhances final product quality.

In certain applications such as royalty measurement of crude oil, accuracy is of paramount importance and virtually no lengths are too great in its pursuit. A word of caution, however: Do not go for high accuracy for its own sake, because it generally increases flowmeter cost and restricts the choice. Assess how important accuracy is and the effect it will have no plant efficiency and product quality—but do not overspecify.

Typical accuracy specifications for the six flowmeters are shown in Table 2-9, but these figures can be denegrated by operating conditions or substandard installations. For example, consider the effect which fluid viscosity has on what are probably the two most accurate flowmeters, the positive displacement meter and the turbine meter: the flow range and linearity of the turbine meter deteriorates with increasing viscosity while the opposite occurs with a positive displacement meter. In a PD meter, performance improves with increasing viscosity, Fig. 2-35.

CALIBRATED ACCURACY

As shown in Fig. 2-36, the PD meter and turbine meter offer the best potential accuracy, but both meters require an initial calibration to eliminate the effect of manufacturing tolerances. Calibrations also may change in service because of wear or damage to moving components. Operating conditions, particularly viscosity, may cause the calibration to change; for optimum accuracy the meters should be calibrated under actual operating conditions at regular intervals.

With an orifice plate, an initial calibration is not required and the discharge coefficient may be derived empirically. Accuracy of an orifice system, however, is considerably inferior to other meter types; this is principally because of the square root relationship between measured head and flow rate and the fact that the accuracy of the differential pressure transmitter is specified as a percentage of fullscale deflection.

A square edged orifice in good condition has a typical accuracy of ±1 percent of reading; the ΔP transmitter increases system inaccuracy by ±0.5 percent fullscale deflection (FSD). Consequently, at full flow rate the orifice system has an overall accuracy of ±1.25 percent that deteriorates to ±7.25 percent of actual flow at 20 percent of maximum flow.

Accuracy of an orifice plate is principally affected by variations in density—a 2 percent density change results in a 1 percent variation in the orifice coefficient. Accuracy also is affected significantly by wear, erosion or damage to the edge of the orifice. For example, 0.01-in. bluntness of a 2-in. bore in a 4-in. pipe increases the coefficient by 2.25 percent. If accuracy is required, a regular check of calibration is essential.

In the majority of cases, the calibration of the variable area flowmeter is predicted from known data and a specific

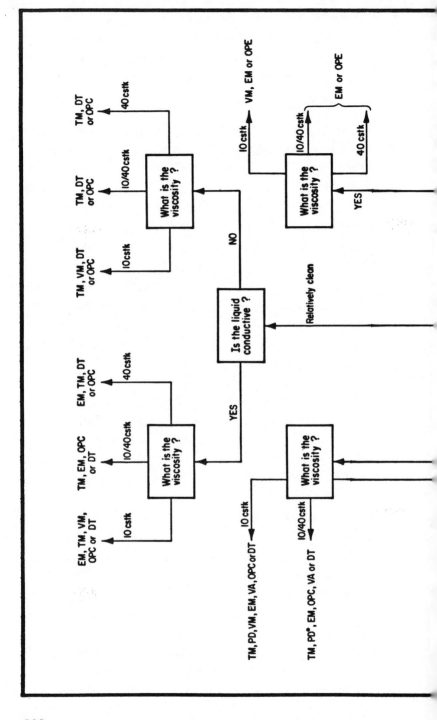

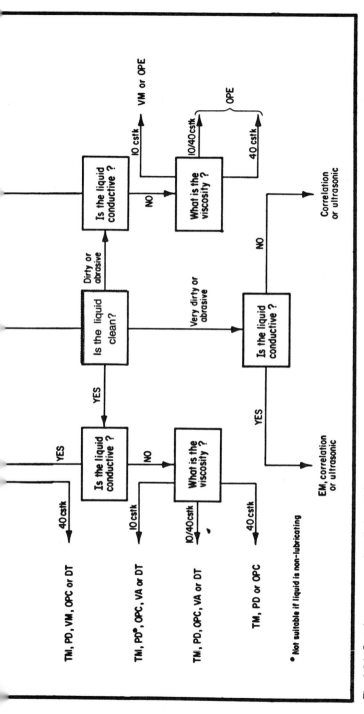

Fig. 2-34. Certain meters can be eliminated from consideration and the choice narrowed down to a reasonable number by following this flowchart. Start in the middle, at "Is the liquid clean?" Abbreviations used are: EM = electromagnetic; DT = Dall tube; TM = turbine meter; VM = vortex meter; PD = positive displacement; VA = variable area of rotameter; OPE = orifice plate, eccentric; and OPC = orifice plate, concentric.

201

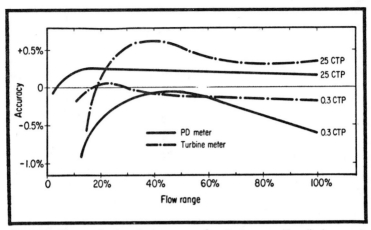

Fig. 2-35. Changes in viscosity have opposite effects on positive displacement and turbine meters. When viscosity increases, positive displacement meters become more accurate; when viscosity decreases, turbine meters are more accurate. Results are for 3-in. meters.

calibration is not performed. To enable the instrument to be used on a variety of different fluids and conditions, either interchangeable scale plates are used or a factor is supplied which converts scale readings to flow rate.

If greater accuracy is required from the variable area meter, then either a "series" check or absolute calibration is performed.

Typical accuracy for an electromagnetic flowmeter is the greater of ±1 percent of actual flow, or ±0.2 percent FSD. At flow ranges above 5:1, the ±0.2 percent FSD accuracy predominates. The effect of fullscale deflection can be eliminated by re-ranging the converter to give a new velocity setting for the 100 percent flow value. Any velocity for fullscale deflection between 2 ft/s and 60 ft/s can be selected at the converter quickly and easily. Consequently, the electromagnetic flow system offers excellent rangeability, and is ideally suited for applications where distinctly different ranges of flow exist, such as day and night plant operations. Accuracy of an electromagnetic flowmeter is independent of operating parameters such as viscosity, density, conductivity, pressure and temperature, but an initial calibration is required.

202

Vortex meter accuracy depends on the operating mode. In frequency mode, accuracy is ±0.5 percent of actual flow rate at all Reynolds numbers above 30,000; but in analog mode, an additional inaccuracy of ±0.2 percent FSD occurs in the frequency-to-analog conversion. Compared with the orifice systems, accuracy of an analog vortex meter is ±0.7 percent fullscale flow rate that reduces to ±1.5 percent of actual flow at 20 percent of maximum flow rate.

Calibration of the vortex meter can be determined from the geometry of the bluff body and an initial wet calibration is not required. Shedding frequency is dependent only on the bluff body size and shape and the associated pipe diameter. With certain bluff body/meter-diameter ratios, manufacturing tolerances have a negligible effect on the meter factor and all

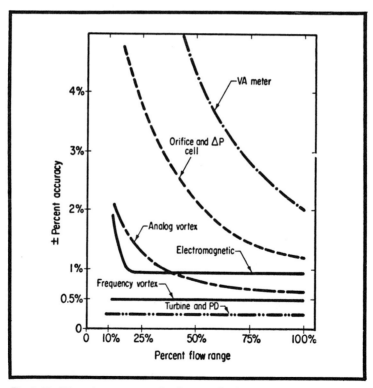

Fig. 2-36. Although accuracy varies from application to application, this chart shows accuracies of the six flowmeters at flow ranges from 0 to 100 percent.

meters of the same size and type have the same meter factor, facilitating interchangeability and system redesign. Calibration is independent of viscosity, density, pressure, temperature; it is also independent of whether the fluid being measured is a gas or a liquid.

With vortex meters, it is not appropriate to quote standard flow ranges—one must talk in terms of "turnup," not "turndown." The limiting factor is the minimum acceptable flow rate, generally set by a Reynolds number of 10,000. The turnup, or flow range, is established by the application's maximum flow rate since the maximum flow rate for the vortex meter is 165 ft/s. Maximum flow can be limited by pressure loss, cavitation (with liquids) or expandability (with gas). With liquids, flow rates of 15:1 are common; with gases, flow rates of 40:1 frequently are possible because of higher pumping velocities.

REPEATABILITY

In many process flow applications, repeatability of a flowmeter is of greater importance than its accuracy. In a flow control loop for example, if the flowmeter gives a stable, repetitive reading, the true accuracy of the measurement is immaterial. The flow value generally can be established or updated empirically on site; and from then on, it is a question of long-term repeatability, rather than sheer accuracy.

It is important to differentiate between long-term and short-term repeatability. Turbine and PD meters offer exceptional short-term repeatability, often as good as 0.02 percent. This characteristic is essential for meter prover duty, but it would not be possible to maintain this level of repeatability over a period of several months. The meter would still give 0.02 percent repeatability, but about a different point.

While the vortex meter cannot match this short-term repeatability, it does offer a very good long-term repeatability of ±0.15 percent. The shedding frequency is a natural phenomenon, and a given size obstruction will produce the same frequency virtually indefinitely.

COMPARING COSTS

Cost is a critical factor in the selection of any equipment. Although it is easy to compare purchase costs in a specific application, it is very difficult to compare costs on a general basis. Variations in pressure and temperature, frequency or analog output requirements, materials of construction, etc. all have significantly different effects on the various meter types.

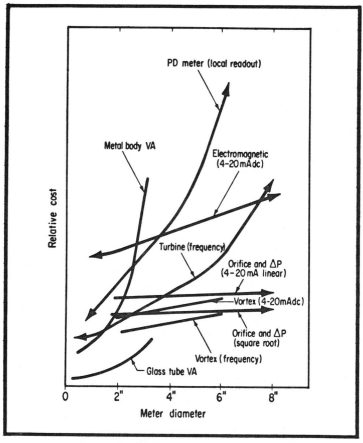

Fig. 2-37. This cost comparison is for a simple water control loop at ambient temperature and low pressure. Standard meter output devices are included in the prices, and the comparison is based on meter diameter, rather than flow rate.

The cost comparisons shown in Fig. 2-37, therefore, are for a simple water control loop at ambient temperature and low pressure. Prices include standard meter outputs (i.e., local readout for a PD or VA meter, 4–20 mAdc for the EM flowmeter, etc.). The comparison is based on meter diameter, but in many cases the meter diameter/flow relationship will vary with meter type. For example, a 6-in. PD meter would be needed to handle the same flow that a 4-in. turbine meter can handle; obviously, this must be considered when comparing costs.

Price comparisons can vary significantly, depending on specific application requirements; but Table 2-10 provides a useful starting point. It is essential, however, that purchase cost not be considered by itself.

When assessing costs, the effects which improved meter accuracy, repeatability or rangeability have on plant efficiency and the end product quality also must be considered. Other factors which should be included on the "balance sheet" are:

- installation
- operating cost
- maintenance
- safety
- pressure loss.

The connection between pressure loss and cost may not be instantly apparent. Consider however the long-term advantages of a low-loss differential pressure meter (see Part I) compared to an orifice plate in terms of decreased pumping requirements. The resulting cost savings are very dramatic in some applications.

INSTALLING AND OPERATING

Installing a meter often can be expensive, inconvenient and sometimes impossible. When problems arise such as straight pipe lengths, providing electrical power or incorporating a filter, installation requirements can be the deciding factor

Table 2-10. Installation Requirements

Installation feature	Orifice plate & transmitter	Electro-magnetic	Positive displacement	Turbine	Vortex	Variable area
Upstream straight pipe	Typically 15/60 D	5 D from electrode	Negligible	10 D	10/30 D, depending on disturbance	Zero
Downstream straight pipe	5 D	Zero	Zero	5 D	5 D	Zero
Mating pipe diameter	Critical for best accuracy	Not critical	Not critical	Critical for best performance	±15%, −5%	Not critical
Power supply	For transmitter	110/240 Vac to detector head	No	No	15/35 Vdc	No
Strainer	Depends on type of orifice plate	No	Yes	Yes	Depends on contamination	No entrained solids
Other features	1. Pressure tapping, manifold and cocks are required. 2. On some applications, gas collection chambers, purging probes, setting chambers, or sealing chambers are required.	1. Use of control flanges for abrasive liquids 2. Ground flanges if pipeline isolated from earth	1. Generally horizontal 2. Pulse generator required for transmission. 3. Special mounting pads required for big meters. 4. Large amount of space required. 5. Installation of large meters is very difficult.	1. Amplifier recommended for long transmission or optimum performance. 2. Flow straightener recommended. 3. Generally requires reduced diameter pipework.	1. Upstream pipe lengths can be reduced if a flow straightener is fitted or if lower accuracy acceptable. 2. Easy installation.	1. Must be mounted vertically. 2. For glass-tube type, fluid must not be opaque.

in flowmeter selection. Even if no problems exist, installation costs should be investigated. Installation costs include materials (pipework, strainers, etc.), labor, and the effect which meeting installation requirements has on system design.

For example, in many applications a PD meter requires a strainer while electromagnetic or vortex meters do not. In this case, the strainer should be included in the cost comparison. An electromagnetic flowmeter without a converter or a turbine meter without a readout unit are nonfunctional; these devices should be included in the cost comparison. The cost of upstream pipework for the orifice place should be compared to the reduced pipe length for a vortex meter or the cost of upstream pipework and flow straightening needed for a turbine meter. Typical installation requirements are summarized in Table 2-10.

After comparing material costs, the next step is to consider installation cost in terms of complexity and manpower required. Does the installation merely involve mounting a pipe spool piece between flanges and taking a pair of wires back to the control room (as in the case of the turbine or vortex meter)? Or, as in the case of an orifice system, does it involve mounting the plate between flanges, providing suitable tappings, fitting a manifold and cocks to the transmitter and then routing a pair of wires back to the control room?

Meter location is also important. A local reading VA meter should be positioned so that there is easy asccess to it, and in a location with good visibility, but where it is not likely to be broken by a blow on the glass tube.

When installation is complete, the flowmeters have to be operated and maintained on a day-to-day basis, adjustments made for process changes and calibrations checked. Because of differences in construction and operation, the flowmeters vary considerably in their operation costs.

Frequency instruments such as the vortex meter or turbine meter generally are operationally simpler than comparable analog units. A frequency system is a go/no-go device which does not require fine tuning, temperature drift compensation or "pot tweaking." An analog unit, however, is less positive and requires checking which in turn, involves "operator repeatability."

The analog output of electromagnetic and vortex flowmeters, however, can be re-ranged quickly on site to give a new value of fullscale deflection; this allows changes in plant operating flow rates to be implemented easily.

MAINTENANCE AND RELIABILITY

Maintenance and reliability are critical factors, which vary among the various meter types and with variations in operating conditions. To quantify operating and maintenance costs the following factors must be considered in relationship to the proposed application:

- labor time/cost involved
- value of components used
- required spares holding
- effect of maintenance downtime on plant operation and efficiency.

The orifice system does not have any moving components, a significant advantage. If accuracy of measurement is required, it is necessary to periodically check the orifice plate for flatness and freedom from buckling and, more particularly, verify that the orifice edge is sharp and square and has not been damaged or subject to any deposition or corrosion.

In addition, it is necessary to ensure that the tappings do not "sludge up" or "gas up." The mainifold and valves must be checked for leakage. At best a leakage is a nuisance and inconvenient, but it also can be dangerous if the metered fluid is either hazardous or obnoxious.

The orifice system has the advantage that the orifice plate can be isolated and the transmitter can be changed without shutting down the process—a significant advantage on some applications.

Excellent performance of the positive displacement meter relies on close tolerance meshing components, which are potentially subject to wear. Regular maintenance and service is therefore advisable, particularly on nonlubricating liquids if the meter's performance is to be maintained. Even on products such as crude oil, the positive displacement meter requires a significantly higher maintenance program than the turbine meter.

Turbine meters have only one moving component—the rotor—and only the bearing is subject to wear. With the use of tungsten carbide for a bearing material, life need not be a problem. A maintenance-free life considerably in excess of seven years can be achieved even on nonlubricating liquids such as butanes and ammonia.

It is essential on a liquid such as ammonia to ensure that it is maintained in liquid form and does not "flash" into gas, otherwise serious meter damage will result. Solids in the liquid can also cause damage to both the turbine and positive displacement meter.

The electromagnetic flowmeter does not have this limitation. In certain applications, electrode fouling can occur unless an appropriate electrode cleaning unit is fitted. EM flowmeters also are available with removable electrodes (see Part I). The modern-day electromagnetic flowmeter does not require any routine maintenance and has a good reliability record. Reliability problems with electromagnetic flowmeters, such as the effect of long-term submersion under water, have been overcome with modern techniques and experience.

The vortex meter does not require any routine maintenance and is an "install and forget" meter. There are no moving, wearing components, and the calibration is far less prone to in-service shift than an orifice plate. The complete electronics can be changed without shutting down the process.

On-line simulator units are available for use with the electromagnetic flowmeter, differential pressure transmitter and vortex meter. Although these units do not verify the true accuracy of the flow calibration, they are invaluable aids in troubleshooting and/or demonstrating the satisfactory operation of a significant part of the system.

After selecting a suitable meter for an application, the meter must be sized and specified to suit the application, installed under acceptable conditions, and then kept at optimum performance through a suitable maintenance program. Guidance in these areas, however, is readily available once the right meter has been selected.

DAVID J. LOMAS is a product specialist engineer with Kent Instruments Ltd., Luton, Bedfordshire, England.

Measuring Flow with Radiotracers

D. F. RHODES

Accurate measurements of flow rate, on line, are important in several refinery and chemical plant operations: material accounting, instrument calibrations and stock loss studies; they also facilitate the solution of difficult process engineering problems. Such measurements can be made by injecting a small quantity of radioactive substance (radiotracer) into a pipeline stream.

This technique has two distinct advantages: the radiotracer is a totally unique material in the system and, the penetration of radiation through piping enables measurements with external detectors which do not interrupt or disturb normal flow conditions. Although the selection of the particular tracer to be injected depends upon a number of different factors, the primary criterion is that it be compatible with the material in the stream. It is important that the tracer be a radionuclide which emits gamma rays having sufficient energy to cause an adequate response in a detector mounted on a pipe. Also, the half-life of the radionuclide should be long enough so as not to pose problems in shipping to remote field locations.

While this method is not generally suitable for continuous flow rate measurements, it provides an accurate and precise means of making an instantaneous measurement for such purposes as the checking of flowmeter calibration (Refs. 2.11—2.19). Since the measurements are not continuous, they are usually performed to troubleshoot a system. Leaks can be located and the flow of material into improper channels can be determined as well. Such testing, done without disturbing normal operation of process units, permits troubles to be located and corrected quickly, thus avoiding unnecessary and costly plant shutdowns. A few of the quantitative measurements made possible by this technique for the solution of process engineering problems include: 1) the degree to which pipeline materials have been mixed, 2) the total volume of material in an open channel having an irregular contour, and 3) the linear velocities of specific components in multiphase systems.

TYPES OF RADIOTRACER MEASUREMENTS

There are several different types of radiotracer measurements which can be used to measure flow rates. Three methods which we have employed are the two-point or pulse velocity method, the total count method, and the continuous sample or isotope dilution method.

Two-point method—A small amount of radiotracer is rapidly injected into the stream; the time of transit t between two detectors, located a known distance apart on a straight uniform section of pipe, is measured. The flow rate is then computed by multiplying the linear velocity by the inside cross-sectional area of the pipe:

$$Q = \frac{\pi d^2 \, l}{4t} \qquad (2.10)$$

where Q is volumetric flow rate; d, the inside diameter of the pipe; and l, the distance between the two detectors. This method has the advantage of being relatively straightforward; however, it is limited in that a length of straight pipe of uniform

214

and known cross-sectional area is necessary. Ideally, this section of pipe should be at least 10 m in length. Obviously the accuracy of the flow rate measured by this procedure is only as good as the accuracy of the available value given for the cross-sectional area of the pipe. API or ASTM pipe standards, under which pipes are normally purchased, permit wall-thickness variations of up to 15 percent on new pipe. Thickness variations can also arise in pipes while in service, due to either scale buildup or corrosion. For example, an unknown scale buildup of 1/32 in. will result in a flow rate error of 3.2 percent on a 4 in. schedule 40-pipe.

Total count method—The total number of counts N is measured and recorded by a pipe-mounted detector during the time that an accurately known amount of radioactivity flows by:

$$N = \int_{t1}^{t2} F \, R_v(t) \, dt \qquad (2.11)$$

where F is a calibration factor and $R_v(t)$ the concentration of radioactivity in the stream as a function of time. After making a change of variable by defining Q as the derivative of volume with respect to time,

$$Q = \frac{dv}{dt} \qquad (2.12)$$

Equation 2 becomes:

$$N = F/Q \int_V R_v(v) \, dv \qquad (2.13)$$

The value of the integral in Equation 2.13 is the summation of radionuclide concentration over the total sample volume, or the total activity injected, A. The calibration factor, F, must be obtained through tests on a piece of pipe having exactly the same geometrical configuration as the process stream. This step in the operation is tedious and can be very time-consuming.

The total count method permits the measurement of flow rates in open streams having an irregular contour, although it can be applied to pipelines as well. It is also useful for leak rate measurements where relative flow rates in different streams are compared. Although this method can provide accurate flow measurements under ideal conditions, it has a number of serious limitations which, in many cases, make it unsuitable.

Results obtained with this method, unlike the two-point method, are not greatly affected by pipe diameter. They are, on the other hand, very sensitive to variations in thickness of the pipe wall; errors as large as 10 percent due to pipe corrosion have been encountered. Substantial fluctuations in background noise due to buildup of radioactivity within the unit or in a nearby storage tank has been a problem, along with electrical noise generated by equipment in the area. In situations where large amounts of insulation are present on high-temperature pipe, it is difficult to determine the calibration factor by accurately reproducing piping geometry.

Continuous sample method—A predetermined amount of radioactivity is injected into the stream and its concentration is examined at a point beyond where mixing takes place. To determine the concentration at the sampling point, a sample is continuously collected while the radiotracer passes. The total activity injected, A, can be expressed by:

$$A = \int_V R_v(v)\, dv \qquad (2.14)$$

After making a change of variable with Equation 2.12:

$$A = Q\int_{t1}^{t2} R_v(t)\, dt \qquad (2.15)$$

The integral in this equation can be replaced by the quantity, $\overline{R}_v\, T$, where $\overline{R}_v$ is the average concentration of radioactivity in the collected sample, and T is the total collection time for the sample. Since the concentration of the tracer is measured only at the sampling point, the samples can be collected in a branch stream which leaves the main pipe after mixing has occurred. This is a useful technique if the main pipe itself is inaccessible.

Besides being totally independent of geometrical characteristics such as pipe diameter and wall thickness, this method offers further advantages. Since a collected sample is counted, the counting time can be made long enough to achieve high statistical precision with only small amounts of tracer being used. Tracer injection and sample collection can be accomplished with relatively simple equipment. A sample can be counted at a remote location, thus eliminating the need to move the counting equipment from one site to another.

One limitation of this method is that suitable injection and collection points must be made available; however, we have found that refinery crews can install a hot tap on a material line in less than one hour. Another limitation is that hot or volatile streams are difficult to handle.

SETTING UP RADIOTRACER TESTS

Figure 2-38 shows a system which employs each of the previously mentioned radiotracer methods. The injection system consists of a 30 ml injection cylinder and a 1-liter chaser cylinder which receives pressure from a nitrogen gas suppply. Although it is not required for the two-point method, this configuration produces optimum results with the continuous sample method.

Each detector contains a sodium iodide (NaI) scintillator activated with thallium, a substance responsive to gamma rays. The 2-in. square scintillator, Fig. 2-39, couples directly to a photomultiplier tube powered by a battery pack, all of which are enclosed in an explosionproof housing. The detectors mounted directly on the pipe, Fig. 2-39A, are used for the two-point and total count method. For the continuous sample method, a similar detector is placed in a counter shield on a tripod stand, Fig. 2-39B, and works on an 8-oz bottle of collected sample.

The dual electronics system for the pipe-mounted detectors, usually installed in a rented panel truck, provides a high degree of reliability, as well as a method of cross-checking results. It consists of two amplifiers with associated single-channel discriminators. The outputs of both detector systems

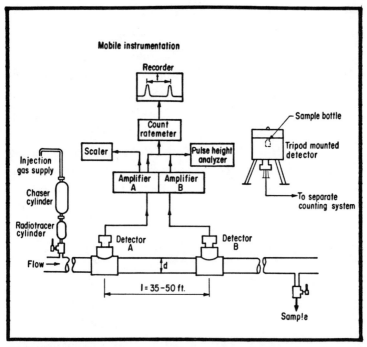

Fig. 2-38. This setup permits flow measurement by any of three methods, after a radionuclide is injected. Two-point: transit time between detectors A and B is measured and linear velocity is multiplied by the cross-sectional area of the pipe; total count: number of counts of radioactivity is detected in a known volume of product passing by detector A (faster flow, fewer counts and lower scaler output); continuous sample: concentration in an accumulated sample is multiplied by the time required to collect all of the injected radioactivity.

are fed simultaneously to a count rate meter and strip chart recorder and to a multichannel pulse height analyzer that drives an output typewriter. The analyzer, operated in the multiscalar mode, makes use of its memory to record counting rate as a function of time.

The technique most frequently used to measure flow rate involves both the two-point method and the continuous sample method. Two detectors are mounted about 35 to 50 ft apart on a straight section of pipe. An accurately measured quantity (5 to 25 ml) of radiotracer is rapidly injected upstream at a suitable distance to permit proper mixing. The transit time of the radiotracer between detectors is recorded by the instruments in the truck. Concurrently, a sample is continuously

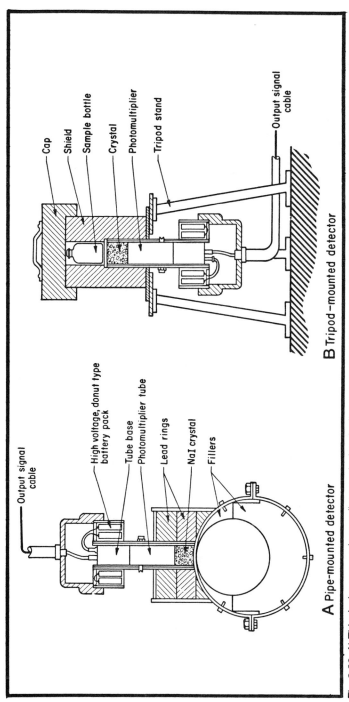

A Pipe-mounted detector

B Tripod-mounted detector

Fig. 2-39. A) This device responds to radiotracer in the pipeline stream by generating an output signal for two-point and total count flow rate measurements. B) This unit, used for the continuous sample method, measures the average concentration of radioactivity in a sample collected in a bottle.

collected at a constant rate at a point beyond the second detector, with collection starting before the tracer reaches the sampling point and continuing until all of the tracer has passed. Collection time is recorded with a stopwatch. This procedure requires a three-man crew; to properly sequence the events, the men communicate by two-way radio.

A series of on-site tests for flowmeter calibration requires about two days' work; testing must be done by personnel experienced in the handling of radioisotopes and the use of nuclear instruments. Several weeks of advance preparation are required to order the radiotracer and to complete any necessary radioactive material licensing arrangements. The cost of radiotracer seldom exceeds a few hundred dollars.

To evaluate the accuracy of the two-point and continuous sample techniques, on-site experiments were conducted at the Gulf Oil refinery in Port Arthur. A series of streams was selected which would permit the greatest number of cross-checks between radiotracer and other methods of flow rate measurement, Streams were also selected for turbulent flow, with Reynolds numbers greater than 10^4. Flow rate measurements obtain with the two radiotracer techniques were compared with readings obtained from on-stream meters as well as tank gage readings. Test results obtained on one of the streams are shown in the Table 2-11.

True values of precision were not readily obtainable from the test data because of the difficulty in distinguishing between variations in flow rate and random fluctuations in measurements. However, relative standard deviations calculated on the basis of daily runs were found to be generally about 2 percent, including fluctuations in actual flow rate.

In single phase liquid streams, flow rate measurements accurate to about 1 percent can be made with the continuous sample method. Flow rates accurate to within 1 percent can also be made with the two-point method in cases where the inside cross-sectional area of pipe is known very accurately.

Table 2-11. Flow Measurement Correlations

Conventional method Meter/tank gage (TG)	Radiotracer methods Two-point (TP)	Continuous sample (CS)	Difference (%) $\left(\dfrac{TP-TG}{TG}\right)100$	$\left(\dfrac{CS-TG}{TG}\right)100$
116.6 bbl/h	116.4 bbl/h		−0.5	
117.0				
117.7				
117.0	115.3	121.9 bbl/h	−1.5	3.0
118.3	116.8	119.9	−1.3	1.2
118.5	116.0	117.0	−2.1	−1.3
118.5	118.1	113.8	−0.3	−1.9
116.0	115.9	117.9	−0.1	0.2
117.7	118.1		0.3	
118.8	119.2		0.3	
116.6	115.3	113.6	−1.1	−2.6
115.2	114.3	116.1	−0.8	0.8
		Series 1 { Average difference[a]	−0.7	0.1
		RMS difference[b]	1.0	1.8
117.9	116.5	117.1	−1.2	−0.7
118.8	116.5	117.1	−1.9	−1.4
118.3	116.5	119.3	−1.5	0.8
117.4	115.9		−1.3	
119.2	116.5		−2.3	
119.7	118.1	114.4	−1.3	−4.4
119.0	115.9	117.4	−2.6	−1.3
119.2	118.1		−0.9	
		Series 2 { Average difference[a]	−1.5	−1.4
		RMS difference[b]	1.6	2.2

221

SOME PRACTICAL APPLICATIONS

Accurate measurement of flow rates for the calibration of flowmeters has been our largest use of radiotracers. To perform these measurements, we have employed both the two-point and the continuous sample methods. A sufficient number of duplicate radiotracer runs were made on each line to insure accurate and reliable determination of meter factors. Cross-checking was done with various radiotracer methods to obtain the highest degree of accuracy. Ultrasonic measurements of the inside diameters of pipes were made whenever possible.

Flowmeter calibrations—At several refineries, comprehensive calibration studies were carried out on the primary flowmeters. We found instances where meter calibrations were in error by as much as 20 percent. After the problem flowmeters were identified, troubles were generally corrected by carrying out mechanical repairs; these were verified by repeated radiotracer tests at a later date.

One rather unusual problem concerned an orifice flowmeter on a butane stream feeding a storage well. During initial tests on this device, a considerable calibration discrepancy was discovered as well as an unexplained nonlinearity; i.e., the meter factor varied as a function of flow rate. After the initial tests, the installation was modified, which improved the calibration considerably. However, a small discrepancy in meter calibration persisted along with a small amount of nonlinearity. We were never able to determine the cause of this nonlinearity; it was not related to compressibility, since we were operating well above saturation pressure.

In another instance, radiotracers were used to check the calibration of a liquefied petroleum gas (LPG) orifice meter used for custody transfer at a terminal. This test was designed to resolve discrepancies between three different methods which were being used to calculate meter coefficient. The radiotracer used for the test was bromine-82 in bromobenzene. Since a sample could not be collected, the two-point technique was employed. Because of the high accuracy needed for this particular measurement, the inside cross-sectional area of the pipe was determined by accurately measuring outside diameter with micrometers and wall thickness with an ultrasonic thickness gauge. Due to the temperature and pressures involved, it was necessary to correct the results with a compressibility factor, to report the meter coefficient at 60 °F saturation pressure. By making these measurements carefully, we were ultimately able to calibrate the meter with an accuracy of better than 0.5 percent.

Leak tests—One important application of radiotracers over the years has been to test for suspected leaks in process equipment. One particular test which was carried out with spectacular results involved the location of a leak in the reactor shroud of a desulfurization unit. During construction, holes had been cut in the reactor shroud to permit the installation and tightening of bolts located underneath. Construction was completed by welding cover plates over these access holes; however in this case, one of the cover plates had been overlooked. This condition caused a significant loss in operating efficiency until the suspected leak was verified by means of a radiotracer test and the appropriate repair work was done.

Other leak-testing applications of radiotracers have involved tests for suspected feed-effluent leaks in heat exchangers of reforming units. In reformer heat exchanger units, the product is cooled by transfer of heat from the effluent into the feed. In the cases we have investigated, the symptoms have been a loss in octane number of the product.

Our procedure for conducting these tests has been to inject a slug of radiotracer into the feed while the radioactivity of the product is monitored. To calibrate the detector and determine timing requirements, radiotracer injections are also made into the line containing product.

The results of our tests on all of these exchangers to date have indicated leak rates of less than 1 percent. Since leak rates of 10 or 20 percent would generally have been required to account for the loss of octane quality, another source had to be causing the difficulty. The solution ultimately involved shutting down the unit for replacement of the catalyst charge.

Water flow tests—Flow rate measurements of water systems are sometimes needed for water conservation studies and also to evaluate heat exchanger performance characteristics. To make this type of measurement, aqueous solutions of bromine-82, iodine-131 and gold-198 radiotracers have been employed. All are commercially available at relatively low cost.

Gas flow tests—Krypton-85 and argon-41 have been used to measure flow rates in gaseous streams. The short half-life of argon-41, 1.8 h, places severe tactical limitations on its use. We have, however, used argon-41 as a radiotracer to measure air velocities through an experimental reactor at our research center, and to measure the linear velocity of the gaseous component in multiphase pipeline streams (Ref. 2.20). Linear velocities as high as 300 ft/s have been measured.

REFERENCES

2.11 Clayton, C. G., "Precise Tracer Measurements of Liquid and Gas Flows," *Nucleonics*, Vol. 18, No. 7, 1960, pp. 96-100.

2.12 Clayton, C. G., "The Measurement of Flow of Liquids and Gases Using Radioactive Isotopes,' *Journal of the British Nuclear Energy Society*, Vol. 3, No. 4, Oct. 1964, pp. 252-268.

2.13 Clayton, C. G., Webb, J. W. and Whittaker, J. B., "The Dispersion of Gas During Turbulent Flow in a Pipe," *British Journal of Applied Physics*, Vol. 14, Nov. 1963, pp. 790-794.

2.14 Ellington, R. T., Staats, W. R. and Kniebes, D. V., "Argon-41 as Gas-Flow Tracers," *Oil & Gas Journal*, Vol. 57, No. 45, Nov. 2, 1959, pp. 99-102.

2.15 Kniebes, D. V., Burket, P.V. and Staats, W. R., "Argon-41 Measures Natural Gas Flow," *Nucleonics*, Vol. 18, No. 6, 1960, pp. 142-147.

2.16 Ljunggren, K., "Review of the Use of Radioactive Tracers for Evaluating Parameters Pertaining to the Flow of Material in Plant and Natural Systems," *Proceedings of the Symposium on Radioisotope Tracers in Industry and Geophysics*, International Atomic Energy Agency, Prague, 1966, pp. 303-348.

2.17 Taylor, G., "The Dispersion of Matter in Turbulent Flow Through a Pipe," *Proceedings of the Royal Society of London*, Vol. A223, 1954, pp. 446-468.

2.18 Hull, D. E., "The Total-Count Technique: A New Principle in Flow Measurement," *International Journal of Applied Radiation & Isotopes*, Vol. 4, 1958, pp. 1-15.

2.19 Ellis, W. R., "A Review of Radioisotope Methods of Stream Gauging," *Journal of Hydrology*, Vol. 5, No. 3, 1967, pp. 233-257.

2.20 McLeod, W. R., Rhodes, D. F. and Day, J. J., "Radiotracers in Gas-Liquid Transportation Problems—A Field Case," *Journal of Petroleum Technology*, Vol. 23, Aug. 1971, pps. 939-947.

DONALD F. RHODES is a Research Physicist in the Nuclear Techniques Group of Gulf Research & Development Co. in Harmarville, PA. Article is based on a paper presented at the 21st Annual ISA Analysis Instrumentation Symposium, King of Prussia, PA, 1975.

Measuring Forces with a Miniature Capacitive Transducer

G.M. BROWN

Based on a single integrated circuit, this transducer responds to variations in pressure or force by changing its output frequency in direct proportion to the displacement of its capacitive sensor. This frequency-modulated device, coupled to either an analog or digital control system, will accurately measure variable-induced displacements of less then five ten-thousandths of an inch, providing adequate sensitivity for force measurements using an elastically deformed member.

Accurate measurements of many variables can be made with capacitive transducers. However, they are usually designed as special-purpose devices, rather than as off-the-shelf commercial items. Until recently, this type of transducer has found priary use in audio work and diaphragm pressure measurements. The proposed design greatly reduces the circuit complexity usually associated with capacitive transducers.

Changes in force or pressure shift the output frequency of this transducer by displacing a parallel plate sensor in the frequency-determining circuit of an integrated circuit function generator. The transducer shifts its frequency output in direct proportion to the amount of displacement of the sensing element. In this configuration, a function generator IC normally

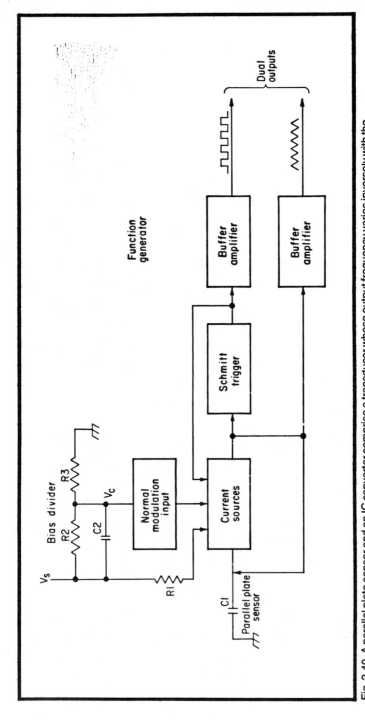

Fig. 2-40. A parallel plate sensor and an IC converter comprise a transducer whose output frequency varies inversely with the RC time constant of R₁ and C₁, the frequency determining components. C₁ changes in value when the distance between its plates changes; thus, output frequency is a function of plate separation.

used as a resistance-controlled or voltage-controlled oscillator serves as a capacitance-to-frequency converter, Figure 2-40. The converter consists of a Signetics SE/NE 566 function generator IC whose characteristics are listed in Table 2-12 and four fixed-value components. The two outputs of the function generator can range up to 1 MHz in frequency. Excluding the sensor, all transducer circuit components could be packaged in a 1-cm cube.

AN FM MEASURING CIRCUIT

Since the capacitance of a parallel plate capacitor varies inversely with the distance between its plates, sensor displacements modulate the frequency of both the square and triangular wave output signals from the converter. One side of the frequency control capacitor is grounded, as shown in the Figure. A resistive divider, R_2R_3, holds the normal modulation

Table 2-12. SE/NE 566 Specifications

Operating characteristic	Typical value
Temperature	70 °C (max)
Supply voltage	24 V (max)
Supply current	7 mA
Operating frequency	1 MHz (max)
Frequency drift	
Temperature	200 ppm/°C
Supply voltage	2%/V
FM distortion	0.2%
(±10% deviation)	
Sweep rate	1 MHz (max)
Sweep range	10:1
Square wave output	
Impedance	50 Ω
Voltage	5.4 V_{pp}
Rise time	20 ns
Fall time	50 ns
Triangle wave output	
Voltage	2.4 V_{pp}
Linearity	0.5%
Impedance	50 Ω

input as a bias voltage V_c that establishes the free-running (unmodulated) output frequency. The bias may be fixed at any level between three-quarters of the supply voltage and the full supply voltage. To eliminate possible oscillation in the control current source, it is recommended that a small capacitor, typically $0.001\ \mu F$, be connected as shown.

Components R_1 and C_1 determine the frequency of the dual output by establishing the RC time constant of the charging path of the current sources. For example, as C_1 decreases in value due to an increase in displacement, less time is necessary for charging through R_1. Thus, frequency varies inversely with the value of C_1 or directly with the amount of displacement.

The output frequency f_o for a given value of C_1 can be approximated by:

$$f_o \simeq \frac{2(V_s - V_c)}{V_s R_1 C_1} \qquad (2.16)$$

where V_s is the supply voltage and V_c is the control bias voltage. Substituting $V_c = kV_s$, the output frequency is:

$$f_o = \frac{2(1-k)}{R_1 C_1} \qquad (2.17)$$

where k is between 0.75 and 1.0, and typical values for R_1 are between $2\ k\Omega$ and $20\ k\Omega$. Since typical capacitance values C_1 are small (less than 200 pF), the output frequeny is generally above 350 kHz.

From Equation 2.17, output frequency is independent of any supply voltage variations. However, some drift actually does occur if the supply voltage varies (about 2 percent frequency shift per volt), as listed in the Table. A 3-terminal IC voltage regulator may be encapsulated with the function generator to eliminate large voltage fluctuations without significantly increasing the physical size of the package.

PARALLEL PLATE SENSOR

The capacitance between the two parallel plates of the displacement sensor can be expressed in picofarads as:

$$C_{(pF)} = 0.00885 \ \epsilon A/d \qquad (2.18)$$

where

ϵ = relative permittivity of dielectric
A = total area of plates in square mm
d = distance between plates in mm.

As an example, a capacitor with a total plate area of 2,500 mm^2 and a separation of 0.5 mm between the plates has a capacitance of 44.25 pF. Assuming typical circuit parameters of 10 V and 7.9 V for the supply and control voltages respectively; and 15 kΩ for resistance R_1 in Equation 2.17, the circuit output frequency is 633 kHz at this capacitance value.

Substituting Equation 2.18 into Equation 2.17, the output frequency as a function of distance between the capacitor plates is:

$$f_o = 22.6 \ (1\text{-}k)\frac{d}{R_1 \epsilon A} \qquad (2.19)$$

Note that if the distance between the plates increases by 0.01 mm (to 0.51 mm) and all other values remain unchanged, the output frequency increases to 646 kHz. The frequency shift of 13 kHz for a displacement of 0.01 mm is readily detectable, and thus the transducer is quite sensitive or displacement.

In some applications, guard rings are installed to minimize fringing effects. Such refinements, however, do not affect the basic response of the parallel plate capacitor.

OUTPUT LINEARITY AND STABILITY

Figure 2-41 is the experimentally determined calibration curve for a parallel plate displacement transducer with a plate area of 3,167 mm^2, and without correction for fringing. Over a

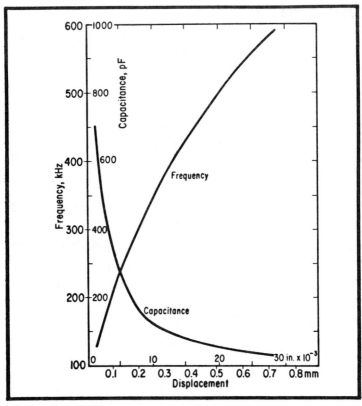

Fig. 2-41. Even though capacitance varies in a nonlinear manner with plate separation, the frequency/displacement relationship is almost linear.

five-hour period, this system remained within 0.1 percent of its initial value. The experimental data show that the frequency/distance relationship is slightly nonlinear; it is, however, a significant improvement over the highly nonlinear capacitance/distance relationship shown in the same Figure. Note also that Equation 2.19 is a reasonable approximation to the actual relationship shown in Fig. 2.42

The frequency modulated output from the capacitance transducer may be transmitted by any of the standard techniques, using either the triangular or square wave output. In conjunction with an output driver and a dual power supply of ±5 V, the transducer can drive standard logic (TTL) circuitry, as shown in Fig. 2-42.

A voltage proportional to the output frequency can be obtained from a phase-locked loop system, such as the Signetics NE 565, which has a lock range of ±60 percent of the free-running frequency. Such a hookup for the transducer whose characteristics are shown in Fig. 2-41 would give a frequency range of 150 kHz to 600 kHz or, in terms of displacement, 0.06 to 0.7 mm. Overall system response may be linearized with a diode function generator or other technique.

TWO PRACTICAL APPLICATIONS

The transducer can be employed in pressure measurements since a movable diaphragm mounted in a metal cell forms a variable capacitor. For small deflections, the capacitance of a solid plate and a clamped diaphragm of radius a, initially separated by a distance d; is:

$$C + \delta C = \frac{\epsilon a^2}{3.6d} + 0.0174 \; \frac{\epsilon(1\text{-}v^2)a^6}{Ed^2t^3} \; p$$

where p is the applied pressure, v is Poisson's ratio, E the elastic modulus, t the diaphragm thickness, and ϵ the permit-

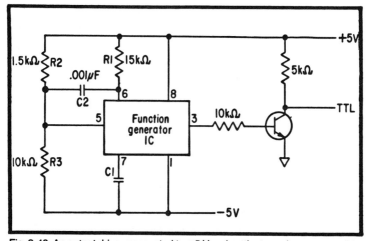

Fig. 2-42. An output driver connected to +5 V makes the transducer compatible with standard logic circuitry. Connections to the SE/NE 566 function generator IC are labeled with pin numbers.

233

tivity of the dielectric. The first term defines the constant capacitance resulting from the initial physical configuration. The fractional capacitance change due to the applied pressure is:

$$\frac{\delta C}{C} = 0.626 \frac{(1-v^2)\, a^4}{E d t^3}\, p$$

A major advantage of this transducer is that it generates an output frequency directly proportional to the applied pressure.

The capacitance-to-frequency converter described here can be also used to construct a zero-interaction biaxial force transducer to measure both tension and torsion simultaneously. Transducers of this type based upon strain gauges are commercially available; however, they usually show a significant degree of interaction between the two force modes. Under test conditions involving large values of one mode combined with small values of the other mode, large percentage errors can occur.

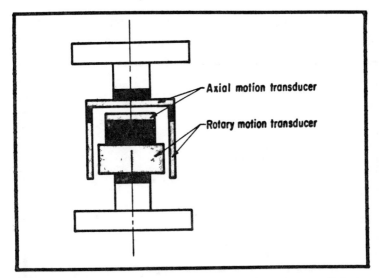

Fig. 2-43. A combination of parallel plate and serrated type sensors eliminates the biaxial interaction that can occur with strain gauge force transducers. Thus, changes in one force mode, either tension or torque, will not affect the measurement of the other force.

234

As an alternate to the strain gauge construction, a parallel plate capacitor can detect deformations due to axial loads. As long as the plates are parallel, rotations of one plate relative to the other will not change the capacitance. The angular deformation due to torsion is relatively small and requires a high-sensitivity rotary transducer to measure it. A serrated capacitor displacement transducer provides the needed combination of high sensitivity to torsional deformation and insensitivity to axial deformation. The capacitance for a serrated-capacitor transducer is:

$$C_{(pF)} = \frac{A}{36\pi t}$$

where A is the overlap area of the elements in square mm and t is the radial gap in mm. Thus, the use of capacitance transducers permits the design of an interaction-free load cell, whose configuration is shown in Fig. 2-43.

GRAHAM M. BROWN is an Assistant Professor in the Department of Materials Engineering at the University of Illinois, Chicago, IL. Article is based on a paper presented at the ISA Aerospace Industries and Test Measurement Divisions Joint Symposium, 1975, Philadelphia.

Errors in Automatic Weigh Filling

V. WHITTAKER

Automatic weigh-filling machines are used in the packaging of a wide variety of products throughout many industries. Their purpose is to dispense materials to a predetermined weight, assuring that packages meet minimum weight requirements and minimizing "giveaway" of materials. No matter how precise the weighing and control system, however, some inaccuracy always exists. This inaccuracy, known as scatter, results from inconsistent variations in density and flow of materials. Inaccuracy exhibits two forms:

- Random fluctuations of individual weights around the average weight
- A shift of the average weight away from the target weight.

Random fluctuations in weight are typically distributed as a bell-shaped curve. If the output weights of a 100-g filling machine are sampled and recorded, 68 percent of the samples usually fall within ± 0.5 g of the target weight. If this is the case, then 95 percent will fall within ± 1.0 g and 99.7 percent with ± 1.5 g. To assure that virtually no underweight fills occur, it is necessary to adjust the average fill of this machine to a value somewhat greater than 1.5 g over the minimum acceptable target weight. Consequently, product giveaway is more than 1.5 percent.

The second type of inaccuracy is a shift of the average weight away from the target value. Random scatter still occurs, but it is centered around a different weight. If this error is likely to occur, it is necessary to set the average target higher so that a combination of scatter and average weight downtrend will not cause any underweight fills.

Inadequate performance of a weigh scale contributes to both types of errors. For example, long-term or thermal instability in a weigh scale causes a shift in the average value; friction of pivots, hysteresis in springs or electronic circuits, or even electrical "noise" contribute to scatter inaccuracy.

Modern electronic weigh scales, especially those using differential transformers as the electrical transducer, provide accuracy and repeatability in excess of the potential accuracy of the total machine system. Their contribution to system inaccuracy is minor. The primary sources of inaccuracy in weigh-filling machines using electronic weigh scales are the mechanics of material handling and variations in material density combined with system delays. Shifts in the average weight value are mainly caused by total system delay, excluding free-fall delay, combined with long-term changes in feed rate or material density.

THE IMPACT OF FREE-FALL

Scatter, or distribution around the average, is caused by several factors. They include nonrepeatability in the delays of the weigh scale, electronics, relays, gate mechanisms, vibratory feeders or motor drives—even hysteresis in the scale/ electronics combination. A predominant factor, however, is the distance of free-fall combined with short-term fluctuations in material flow.

The extent of inaccuracy caused by very short-term variations in material flow rate is not obvious. Packagers generally assume that fluctuations in rate of feed simply cause

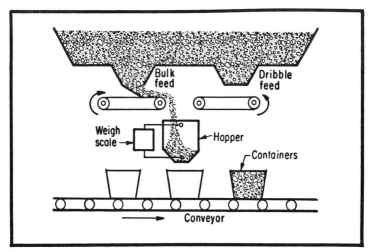

Fig. 2-44. Weigh-filling machine designed to dispense material in bulk and dribble feed.

equivalent fluctuations of weight received at the weigh scale, but this is not the case. In reality, fluctuations in feed rate cause gross error signals and the magnitude of the error is related to the distance of free-fall.

Consider the dynamics of material flow in a filling machine, Fig. 2-44. This machine dispenses 450 g into an intermediate hopper in 3 seconds. Target weight is achieved using two flow rates: a bulk flow of 270 g/s and a dribble flow of 30 g/s. Based on 90 percent of target weight, bulk flow is interrupted at a weight of 405 g on the scale. The dribble is interrupted at 450 g. Mechanical considerations make it necessary to provide space between the hopper and the feeder, resulting in a free-fall distance of approximately 8 in. To simplify the analysis, the free-fall distance is considered constant throughout the fill cycle.

When the cycle is initiated, material is dispensed from the feeder at the bulk rate and falls into the hopper. Figure 2-45 illustrates the material flow from the feeder (solid line) and the resultant weight force on the weigh scale (dotted line). There is no change of weight at the weigh scale for the initial period of time (0.2 s) required for the material to fall 8 in. When the material stream impacts on the hopper surface, no significant weight has accumulated. The impact force, however, causes the weigh scale to react.

The applied impact force is equal to the rate of change of momentum of the material stream, and is equal to rv/g weight units; where r is the flow rate in weight units per second, and v is the velocity at impact. If v is the velocity resulting from free-fall under gravity for a time period t_1, then v equals gt_1. By substitution, rate of change of momentum equals rgt_1/g, or rt_1 weight units. Note that this force is equal to the weight of material dispensed in the free-fall time period—in other words, the weight of material in suspension.

OVERSHOOTING THE TARGET

At the instant of the material impact, the weight scale signal assumes a value equal to the weight of material dispensed by the feeder in the free-fall time period. After the impact force is applied to the weigh scale, the weight increases at the rate material is being dispensed. At time t_2, Fig. 2-45, material flow is hanged to dribble rate. The weigh scale does not sense this occurrence for a time period t_1, since material arriving at the weigh scale through this period was dispensed at the bulk rate. As a result, the weigh scale signal overshoots the weight of material actually dispensed. At time $(t_2 + t_1)$, material impacting the weigh scale is abruptly changed to the dribble rate, and there is an immediate change in the scale signal to a lower weight equal to material dispensed during the bulk cycle plus the momentum signal of the dribble rate.

At time t_3, dribble cutoff occurs, causing a similar overshoot of the scale signal for a period t_1. This overshoot is of smaller magnitude than the experienced with bulk rate flow cutoff. Note that, for a time period t_1 after bulk flow is discontinued, a considerable error appears on the weigh scale. In fact, in this example, the overshoot reaches a weight of 459 g—sufficient to prematurely cut off the dribble feed unless this is prevented by suitable design of the control logic. Figure

240

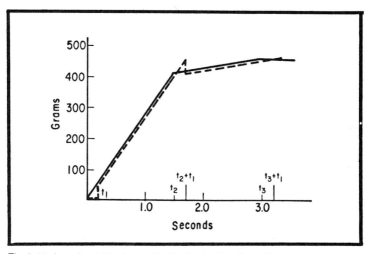

Fig. 2-45. Actual weight released by the feeder is indicated by the solid line. The dotted line indicates weight recorded by the scale. Except for initial startup and bulk cutoff periods, indicated weight for a constant flow rate is equal to weight dispensed because the impact force on the scale is equal to the weight of material in free-fall. When the feed rate is changed from bulk to dribble, the scale momentarily overshoots by a value equal to the impact force at the time of bulk cutoff.

2-45 graphically demonstrates that the weight force applied to the weigh scale of a filling machine, at any instant is equal to the weight of material dispensed by the feeder up to a time preceding that instant by t_1 (the free-fall time), plus a force equal to a projection of the flow rate for a time t_1.

FLUCTUATIONS IN MATERIAL FLOW

Fluctuations in feed rate or nonuniformity in density cause scatter of the weights. The effect of these fluctuations, even with short duration flow discontinuities, is readily apparent in Figs. 2-46 and 2-47. Figure 2-46 illustrates the example of a sudden increase in flow rate, possibly caused by bunching of materials. The weigh scale does not detect the occurrence until some time after it actually occurs—the free-fall time, over which period large transient force is applied to the weigh scale. The magnitude of the transient force increases proportionally with increases in free-fall time. Figure 2-47 shows the converse situation where there is a sudden reduction in flow—possibly caused by a gap in the material stream. In each case such transient forces, occurring close to the final cutoff point, can result in errors.

SELECTING A SCALE

The weigh scale in a filling machine is subjected to both weight dispensed and force from the free-fall. The force component is proportional to the rate of material flow, and to the time of free-fall. Any changes in rate of flow will generate an error proportional to the change in rate and to the time of free-fall. Weigh-filling machine designers can minimize scatter from the free-fall effect by observing three design parameters:

- Maintain constant flow rates with greatest possible accuracy

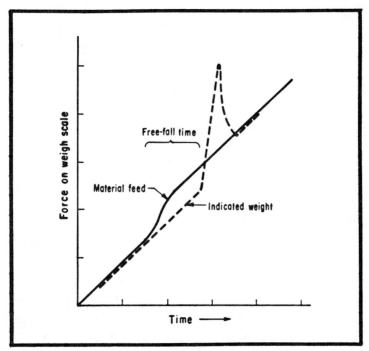

Fig. 2-46. A sudden increase in feed rate is not detected by the scale until after the free-fall time. The impact of the material causes a large upward transient. If this situation occurs near the target weight, it could cause premature cutoff.

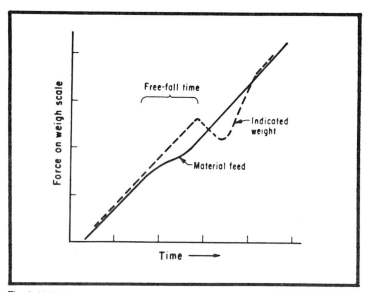

Fig. 2-47. A decrease in feed rate has a opposite effect on the scale, causing a large downward transient. If this situation occurs near the target weight, it could cause an overfill.

- Minimize the distance of free-fall
- Use a weigh scale fast enough to avoid serious system delays, but with insufficient speed to respond completely to transient and rapid flow rate fluctuations.

It is apparent from the third requirement that the choice of weigh scale for a filling machine is always a compromise. If the scale is slow in response there will be a significant system delay and the cutoff point must anticipate the delay; this will result in error from feed rate or density changes. If the scale is very fast, it will respond to transients generated by flow rate fluctuations and scatter will result. A weigh scale, such as a spring-type differential transformer weigh cell or load cell, with a natural resonant frequency between 5 cycles per second and 10 cycles per second, generally offers the optimum performance.

VERNON WHITTAKER is product manager for Automatic Timing & Controls, Inc., King of Prussia, PA.

Measuring Torsional Vibration

V. H. VERHOEF

Torsional vibration is present in all rotating mechanisms, but it is not as easily detected or measured as is linear vibration. Torsional vibration is the periodic change in angular position, velocity or acceleration of a seismic mass. Excessive torsional vibration can cause fatigue failures or shaft fractures, and it can prematurely wear out couplings, gears and bearings. In reciprocating engines, it can cause excessive linear vibration and objectionable noises from banging couplings, gears and bearings.

The amplitude θ of a pure sinusoidal torsional vibration can be expressed as $\theta\sin\omega t$ in radians or degrees. Velocity, then, is $\omega\theta\cos\omega t$ and angular acceleration is $-\omega^2\theta\sin\omega t$, where

θ = peak amplitude in radians or degrees
$\omega = 2\pi$ times vibration frequency

The ω^2 factor can produce very high angular accelerations at high frequencies. High torques result when a large mass is involved because torque is equal to the angular acceleration times the mass moment of inertia.

Problems occur when the torsional vibration forced upon a shaft-mass system is near (or on) one of the system's resonant frequencies. When the system vibrates at a resonant frequency, it is said to be operating at a critical speed. At critical speed, a small amount of torsional vibration is greatly amplified, and the amplification is limited only by the damping or friction present in the shaft material, bearings or couplings.

In a single shaft-mass system, Fig. 2-48A, the resonant frequency in hertz is:

$$f_n = \frac{1}{2\pi} \sqrt{\frac{K}{I}}$$

where K is shaft torsional stiffness in lb-in./radian and I is the mass moment of inertia in lb-in.-s.

For two masses connected by a shaft, Fig. 2-48B, the resonant frequency is:

$$f_n = \frac{1}{2\pi} \sqrt{\frac{K\,(I_L + I_S)}{I_L\,I_S}}$$

When many masses and shafts are involved, finding resonance frequencies becomes much more complex, requiring the use of graphs, tables or a digital computer.

CAUSES AND CURES

Torsional vibration can occur because of loose or worn components, or because of inherent characteristics in a particular driving force. Reciprocating engines or pumps, for example, provide torque that is far from smooth. The varying torque components in a single-cylinder four-cycle engine cause resonant frequencies at 0.5, 1, 1.5, 2, 2.5 (and so on) times the engine rpm; a six-cylinder engine with 120° crankshaft spacings has resonant frequencies at 3, 6, 9, 12, etc., times the engine rpm.

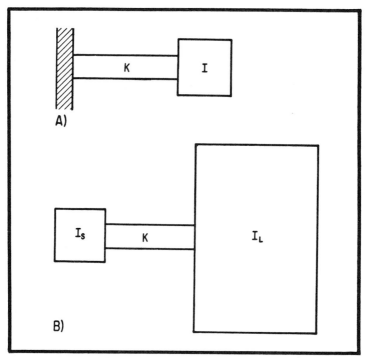

Fig. 2-48. The resonance frequency of a single shaft-mass system (A) or a two-mass system (B) is a function of shaft stiffness K and the moment or moments of inertia I.

Electric motors also do not provide smooth torque; the torque component of an electric motor varies at a frequency equal to twice the line frequency. Torsional vibration is quite noticeable under high load conditions, such as motor startup.

Gear and pulley eccentricity, tooth-to-tooth errors, loosely mating gears and worn gears can cause torsional vibration. Backlash in loose couplings and bearings with high friction spots can cause problems. Other sources include cam shafts, slipping belts, gear-driven pumps, detent mechanisms, stepper motors, and pulsing loads on propellers and fans caused by blades passing fixed objects. Universal joints are well-known for creating second harmonic (rpm × 2) vibrations; in fact, they create the second harmonic so reliably, U-joints are used to calibrate torsional vibration sensors, Fig. 2-49.

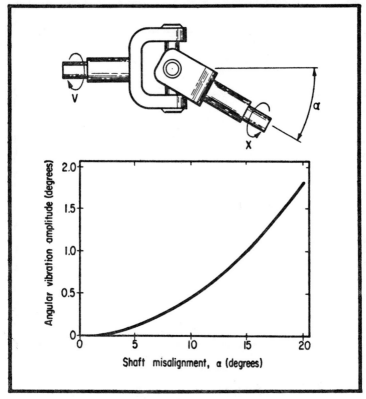

Fig. 2-49. A rotating U-joint creates second harmonic torsional vibration so reliably that it can be used to calibrate transducers. The peak vibration amplitude θ is a function of input and output shaft misalignment angle α.

A flywheel can be used to cure many fibration problems. Although flywheels primarily are used to store kinetic energy, they also reduce the amplitude of torsional vibration because $\theta = T/I$; i.e., the amplitude θ reduces when the moment of inertia I increases.

Other remedies include:

- more cylinders in an internal combustion reciprocating engine
- change of shaft stiffness to avoid resonance frequencies
- tighter couplings
- free-running bearings.

Dampers can also be used to reduce vibrations. A damper consists of a seismic mass coupled to the shaft by dry or fluid friction. Another type uses tuned pendulum masses to dampen a particular resonant frequency.

MEASUREMENT TECHNIQUES

Various ingenious methods exist for measuring torsional vibration. Most mechanical and electrical vibration transducers use a spring-mass system, Fig. 2-50, which has a very low torsional resonance frequency. The spring can be a flexible torsion bar, a helix spring or the attraction force of a permanent magnet.

If the vibration frequency of the input shaft is higher than the resonance frequency of the transducer's spring-mass system, the mass will become isolated from the torsional vibration. The mass will "stand still" under such conditions and can be used as a reference point for measuring vibration amplitude.

Since many of these transducers rotate with the shaft on which they are mounted, slip rings are used to transmit the transducer signal; other types of transducers have a stationary sensing circuit that does not require slip rings.

For measuring the relative motion between the input shaft and the mass, the same principles used in linear dis-

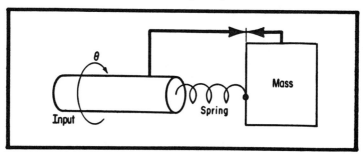

Fig. 2-50. In a spring-mass transducer, the input shaft and seismic mass are connected by a torsion spring. If pointers were to be attached on both the mass and shaft, the relative motion between the pointers would be proportional to torsional vibration. In actual transducers, the springs can be torsion bars, helical springs or magnets, and the "pointers" can be coils, inductive pickups and the like.

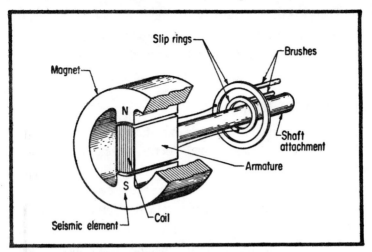

Fig. 2-51. In an inductive transducer, a ball bearing-mounted magnet acts as a seismic mass, and its magnetic field acts as a spring. Torsional vibration causes the coil to move in the magnetic field, resulting in output voltages that are proportional to angular velocity.

placement transducers are employed. Typical methods include self-generating inductive, linear variable differential transformer (LVDT), strain gauge, capacitive, piezoelectric, visible light and eddy-current proximity sensors.

SPRING-MASS SENSORS

An inductive pickup built by the CEC Div. of Bell & Howell, Fig. 2-51, is a good example of a spring-mass transducer that uses slip rings. The mass is mounted on ball bearings placed around an armature and the armature is attached to the input shaft. The strong magnetic field, in addition to acting as a spring, generates a voltage when torsional vibration moves the coil into the magnetic field. Output of the sensor is proportional to the angular velocity. This particular sensor has a sensitivity of 9 mV/deg/s, a frequency range from 10 to 10,000 Hz, and a maximum amplitude of 2 deg.

Another inductive spring-mass pickup, Fig. 2-52, can be used to measure torsional vibration in shafts "without ends," such as the shaft between a propeller and an engine in a ship. The hand-held sensor is built by the Institute TNO (Toegepast

Naturwetenschappelijk Onderzoek) in the Netherlands, and features sensitivity of 160 mV/rad/s and a maximum frequency of 75 Hz; the frequency is dependent upon the properties of the rubber O-rings and the contact pressure.

STRAIN GAUGE TRANSDUCERS

An interesting vibration transducer has been developed by David E. Hamann at Oregon State Univesity. The sensor, Fig. 2-52, contains two cantilever leaf springs and four strain

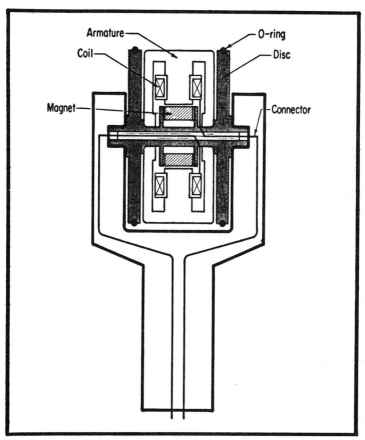

Fig. 2-52. This hand-held device is pressed against a rotating member, such as a driveshaft, to measure torsional vibration. The two discs have rubber O-rings which contact the rotating member, transferring rotational motion to the magnets. The armature and coils form a springmass transducer system. Sensor output is transmitted through connectors in the handle.

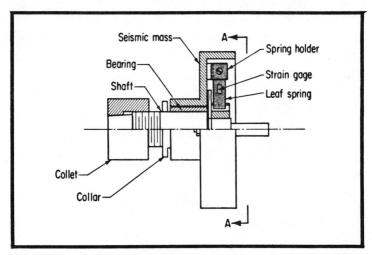

Fig. 2-53. Four strain gauges mounted on two cantilever leaf springs serve as the sensor in this vibration transducer. Relative motion between the input shaft and the seismic mass deflects the springs, and the resulting strain gauge output is proportional to the deflection.

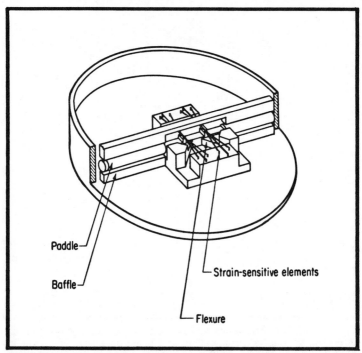

Fig. 2-54. A paddle suspended by strain-sensitive wires forms an unbonded strain gage that responds to torsional vibration. The device is built by Statham Laboratories.

gauges that form a spring connection between the seismic mass and the input shaft. Relative motion between the mass and shaft deflects the spring and causes an output from the strain gauges.

Output voltage, processed by a Wheatstone bridge, is proportional to the angular displacement of the input shaft. Callibration can be done by holding the mass and turning the shaft a known amount. The unit has an output sensitivity of 5mV/deg with a 10-V supply, resonant frequency is 7 Hz and its usable range is from 15 to over 110 Hz.

Statham Laboratories has developed an angular accelerometer, Fig. 2-54, that can operate near its own resonance frequency. A paddle is suspended by strain-sensitive wires that form an unbonded strain gauge. The housing is filled with a fluid that acts as a seismic mass, and the unbonded strain gauge serves as a spring and sensor. Fluid viscosity and the gap between the paddle and the baffles provide damping.

TRANSDUCERS SANS SLIP RINGS

The transducer shown in Fig. 2-55 uses two stationary magnetic pickups at opposite ends of a "torsion shaft" to measure angular deflection. Each encoder detects the pas-

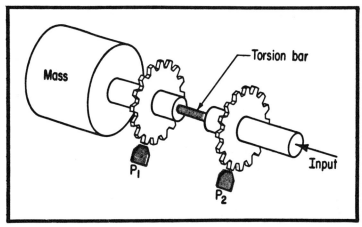

Fig. 2-55. The twisting motion of the torsion bar caused by torsional vibration changes the time between the magnetic pickup pulses in this transducer.

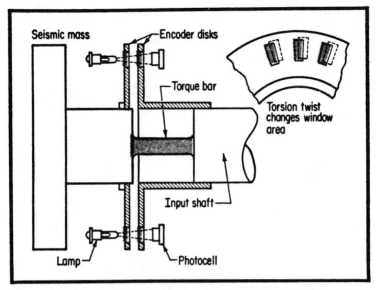

Fig. 2-56. Torsional vibration twists a shaft in this Vibrac Corp. transducer, changing the lineup of windows on encoder discs. The resulting change in light intensity is recorded by photocells.

sage of a gear and outputs a pulse to a flip-flop. The resonance frequency of the seismic mass and the torsion bar determine whether the transducer measures angular acceleration or displacement. Under torsional vibration, the torsion shaft will twist and change the time between pulses. Pulse outputs are sent to a filter which averages the output signal, removes the shaft speed component, and presents a torsional vibration signal to an oscilloscope.

A variation of this technique, Fig. 2-56, uses photoelectric cells to measure the rotation between two discs. Each disc contains windows which, under no-load conditions, line up exactly. As the torque load increases, the shaft deflects and changes window alignment. The change in light transmission from the lamp to the photocell is proportional to the angular deflection of the discs caused by torsional vibration.

ESCHEWING SPRING-MASS

A few torsional vibration sensors exist which do not use a spring-mass system. An eddy-current sensor built by Hood-

254

win Instruments, Fig. 2-57, uses a metal disc, magnet and coil as the sensing element. The disc, which is connected to the input shaft, rotates near a permanent magnet which generates eddy currents in the disc. A pickup coil on the othe side of the disc senses changes in these currents produced by torsional vibrations.

The output voltage of this sensor is proportional to the angular acceleration of the input shaft. The sensor's sensitivity ranges from 0.1 to 10 mV/rad/s^2, depending on the disc material and type of pickup coil. Resonance frequency varies from 800 to 3,000 Hz and maximum rpm is limited by heat generated in the disc. Disc speed limits are 500 to 5,800 rpm, depending on the cooling method and materials.

Another no-mass sensor uses sine waves recorded on a magnetic tape to detect torsional vibration. The system, Fig. 2-58, consists of a magnetic tape attached to or wrapped around a shaft, an erase head, read head and write head. As the tape passes the first head, its contents are erased; next, a write head records a sinusoidal signal which is detected by the read head. The phase angle between the write and read signals is a function of tape speed, and the circuit shown in Fig.

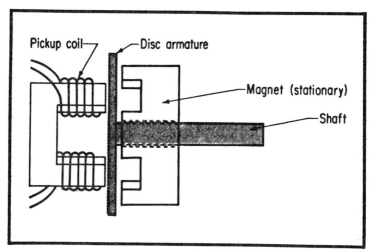

Fig. 2-57. Torsional vibration changes the eddy currents generated by the disc armature and magnet; the eddy-current changes are detected by a pickup coil.

2-58 converts this phase shift to a voltage proportional to torsional vibration. If the bandwidth of the tape heads is wide enough, pulse-type signals can be used, eliminating the need for pulse-generating circuitry.

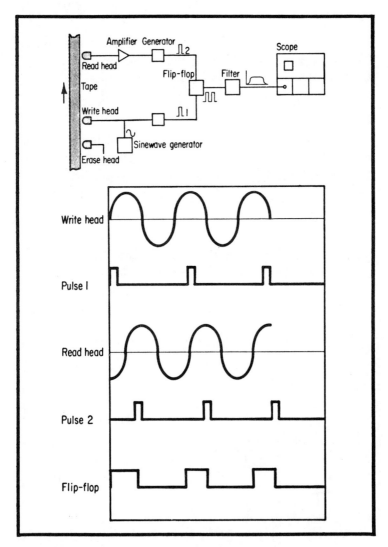

Fig. 2-58. Writing and then reading a sinusoidal wave pattern on magnetic tape is a technique most often used for detecting flutter in a tape transport, but it can also be used to detect torsional vibration. Circuitry detects the phase change in each signal; generating a pulse that drives a flip-flop. After passing through a filter, the signal can be observed on an oscilloscope.

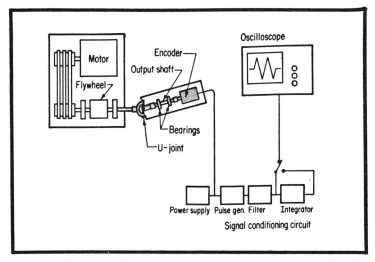

Fig. 2-59. Using an ordinary photoelectric position encoder, the author developed this interesting transducer that outputs both rotational speed and torsional vibration. After passing through a pulse generator, filter nd integrator, the AC component of the signal represents vibration and the DC component is rotational speed.

ENCODER SENSOR

A standard photoelectric shaft-position encoder, commonly used on numerically controlled machines, can be adapted to sense shaft speed and torsional vibration. Figure 2-59 shows the encoder connected via a U-joint to the output shaft of a flywheel. Outputs from the encoder trigger a pulse generator which produces constant time duration pulses. Additional circuitry converts the pulses into an average voltage that represents shaft speed.

The voltage signal has two components: the AC component is proportional to the torsional vibration, and the DC component is proportional to shaft speed. Either component can be monitored by switching the oscilloscope to the AC or DC input position.

Using the system shown in Fig. 2-59, an encoder with an output of 500 pulses per revolution and a generator that outputs 5V pulses of 20 μs duration produced a sensitivity of one mV/deg of torsional vibration and 0.8 mV/rpm for shaft speed. The filter determines frequency limits of the system; with 7.5 kΩ resistors, the output is proportional to angular vibration between 3 and 100 Hz. Replacing the resistors with 1 kΩ components would bring the maximum frequency to 750 Hz, but it would also increase the signal-to-noise ratio.

Maximum amplitude is limited only by the acceleration capabilities of the encoder, and the pulse frequency limitation of the encoder determines maximum rpm. With 500 pulses per revolution, the encoder is capable of 10,000 rpm.

WILLIAM H. VERHOEF is a project engineer at Tektronix, Inc., Beaverton, OR. Article is based on a paper given at the ISA Aerospace and Test Measurement Divisions symposium, Las Vegas, 1977.

Section III—Control Valves

Control valves are as old as industrial processes and as new as the latest computer program. They perform tasks both common and exotic.

"Control Valve Selection" reviews the logic chain and decision techniques by which the instrument engineer makes his final selection on a control valve for a process-industry application. This introduction establishes an overall look at valves, and the next three selections investigate particular areas of valves.

For example, "Control Valve Dynamics" explores the changing forces at work and offers techniques for configuring an actuator to a globe valve.

In "Characterized Valve Actuators," the author offers a method of obtaining improved operation from valves. Benefits can include improved controllability; rangeability, durability, maintainability, cost, or a combination of these factors.

"Predicting Control Valve Noise from Pipe Vibrations," describes a practical method of calculating the noise generated by a valve without resorting to test chambers or having to shut down the rest of the piping system. It describes a method of using the excitation of the pipe wall to calculate valve noise. High ambient sound levels in the vicinity of a valve do not have to impede the measurement of valve noise.

Control Valves

D.G. WOLTER

The most common final control element is the control valve, whose application to industrial processes has advanced from a beginning in the 1930s, to an art in the 1950s, and to a near science in the 1970s. Selection of the proper valve requires consideration of many factors to meet the demands of the process and provide the required mechanical reliability and control performance.

Many kinds of control valves are on the market today to meet routine and unusual applications. New processes with more severe operating requirements and stricter environmental codes dictate the need for special and advanced designs. On the manufacturer's side, mass production economies pressure the suppliers to produce standardized designs that can meet most applications.

The burden is on the *specifying instrument engineer* to bring together all essential requirements, and to apply premium cost features when—and only when—they are needed.

PROCESS REQUIREMENTS

The process description is usually presented to the instrument engineer in the form of a process flow diagram that defines the design flow rates, pressures, temperatures and perhaps some physical data on the process stream at each stage in the process. This diagram then becomes the basis for a mechanical flow diagram, a one-line description of the process with the addition of mechanical requirements. Material selections, pump specifications, equipment sizes, instrumentation and valving requirements are typical of the data shown.

Thus, with the information presented on the process and mechanical flow diagrams, the engineer has a start on the design information required for application and selection of control valves. This data may not cover a good many physical properties (or such expected variances as the minimum flow and the maximum flow), so additional information often must be obtained. The instrument engineer must understand the process in its normal state, at startup and during any abnormal situations—before he makes the valve selections.

At this early point in the selection process, it is time to record data on the control valve specification sheet. A readily available document, ISA Form S20.50, can become a checklist for critical application data upon which a selection will ultimately be based, Fig. 3-1. Unfortunately, this form cannot help in the selection of control characteristics; but it does adequately cover process and mechanical aspects and eases the selection task because of its organization.

Some critical assessments can be made from the fluid description and the available pressure drop. These define service severity and go a long way toward valve selection. For example, a liquid at low inlet and outlet pressures indicates the economical choice of butterfly, ball or globe valve—all with fairly standard designs.

On the other hand, a high-pressure gas with an outlet pressure less than about one-half of the absolute inlet pressure signals the possibility of a noise generation problem, and further investigation is required. Similarly, one must be alert for liquids with vapor pressures near the control valve outlet

GENERAL	1.	Tag No.			
	2.	Service			
	3.	Line No./Vessel No.			
	4.	Line Size/Sched. No.			
BODY	5.	Type of Body			
	6.	Body Size	Port Size		
	7.	Guiding	No. of Ports		
	8.	End Conn. & Rating			
	9.	Body Material			
	10.	Packing Material			
	11.	Lubricator	Isolating Valve		
	12.	Bonnet Type			
	13.	Trim Form			
	14.	Trim Material	Seat/Plug		
			Shaft Mtl.		
	15.	Required Seat Tightness			
	16.	Max. Allow., Sound Level dBA			
ACTUATOR	17.	Model No. & Size			
	18.	Type of Actuator			
	19.	Close at	Open at		
	20.	Flow Action to			
	21.	Fail Position			
	22.	Handwheel & Location			
POSIT.	23.	MFR. & Model No.			
	24.	Filt. Reg.	Gages	Bypass	
	25.	Input Signal			
	26.	Output Signal			
	27.	Air Supply Pressure			
TRANSDUCER	28.	Make & Model No.			
	29.	Input Signal			
	30.	Output Signal			
OPTIONS	31.				
	32.				
	33.				
SERVICE	34.	FLOW UNITS			
	35.	Fluid			
	36.	Quant. Max.	C_V		
	37.	Quant. Oper.	C_V		
	38.	Valve C_V	Valve F_L		
	39.	Norm. Inlet Press.	ΔP		
	40.	Max. Inlet Press.			
	41.	Max. Shut Off	ΔP		
	42.	Temp. Max.	Operating		
	43.	Oper. sp. gr.	Mol. Wt.		
	44.	Oper. Visc.	% Flash		
	45.	% Superheat	% Solids		
	46.	Vapor Press.	Crit. Press.		
	47.	Predicted Sound Level dBA			
	48.	Manufacturer			
	49.	Model No.			
Notes:					

Fig. 3-1. ISA form S20.50 is used as shown—or in some variation—by many engineers for the specification of control valves.

pressure, a condition indicating potential cavitation or flashing and the need for corrective measures.

Other process properties must also be examined. Fluid cleanliness needs review to assure that a valve adequately meets the challenge of entrained solids, slurries, wax-like materials that may cause coating or sticking and any other stream properties which may impair valve performance. If the valve must be flushed clean of contaminants or product residuals before opening for maintenance, or for cleaning at the end of the process period, then it must offer a free-flow path to facilitate cleaning. Freezing possibilities must be reviewed because the flowing material may have a pour point well above prevailing ambient temperatures. Heat tracing or jacketing may be required.

Viscosity can be a serious problem, especially when the fluid is non-Newtonian (e.g., high viscosity gels, colloids, slurries or polymers) or if the Reynolds number of the fluid passing through the valve is below 33,000. In these cases it is necesary to calculate a correction factor for C_v determination (Ref. 3.1).

Shutoff requirements may be critical to the operation of the process. Can sizable seat leakage be tolerated, or is absolute tight shutoff required? Tight shutoff is often needed to prevent process materials from migrating downstream during emergency shutdowns—or even during some normal operating conditions. This consideration is too often overlooked in the valve selection process, but it greatly influences the type of valve that should be employed.

ISA RP39.6 now defines leakage rates from Class I, corresponding to an unguaranteed rate of 0.01 to 0.05 percent of valve C_v, though Class VI. The latter deals with leakage flow of "bubbles per minute" of air at specified test differential pressures and is applicable to resilient-seated valves. This definition eliminates any need for the uncertainty caused by such terms as tight shutoff, metal-to-metal seating, and bubble-tight shutoff.

Process considerations will dictate some specific mechanical requirements within the valve. Process and mechanical parameters often merge and complement each other to help define the application further.

MECHANICAL DECISIONS

Mechanical considerations in control valve selection are probably the easiest to deal with. As plant design progresses toward the point that control valves are selected, mechanical design groups have already established the pressure and temperature ratings the configuration of the piping and probably even provided some guidance on metallurgy requirements for the control valve.

The choice of valve materials is often based on previous experience or on guidance from materials specialists. The instrument engineer must be cautious about sea and plug materials to assure that they are sufficiently corrosion-resistant and hard enough to withstand erosion and wire drawing that may occur because of the pressure drop. The seat ring, plug and stem are normally made from materials slightly superior to the valve body; these parts are subjected to high wear and are more critical in their dimensional integrity (i.e., control characteristics and shutoff leakage capability).

In the control loop, the control valve is an instrument component that is in contact with the process. A transmitter, controller and other items can be removed—or bypassed—for short time periods and maintained as required. Unless special designs are considered, the control valve cannot be removed. The instrument engineer must assure that the plant integrity is properly protected with appropriate maintenance facilities to permit plant operation within the desired program.

If the plant can be shut down easily, then it may not be necessary to include special facilities. If the plant can be placed in a recycle operation, or otherwise idled, it may be sufficient to provide only block and drain valves for quick removal of the control valve. However, if the valve is essential to continuous operation, then it will be necessary to install a full bypass and block valve manifold to permit continued manual operation while the valve is removed for maintenance. It is important to review maintenance strategy at the earliest possible time in plant design: The ramifications affect piping design, plant cost and application of other equipment.

There are valves designed to facilitate in-place maintenance and repair. Manufacturers often highlight this design capability, but personal experience shows that there is very little maintainence—other than packing adjustments or lubrication—which can be accomplished in the field. In addition, most of these designs incorporate additional seals that are prone to leak and that contribute to overall maintenance requirements. While it is easy to support the economics of in-line repair, most maintenance personnel offer convincing arguments that such repairs are usually only partially successful and eventually the valve must be removed for complete overhaul in a shop.

Selection of control valves should also include consideration of spare parts requirements. Purchases should be made to minimize the variety of valves and thereby minimize the spare parts requirements. In addition to the parts advantage, this policy provides an added benefit in that maintenance and operations personnel have fewer valve types to get acquainted with.

The valve manufacturers can assist with the mechanical selection and their expertise should be utilized to the fullest. It may be necessary to modify their recommendations to be more conservative (or less), for example, when it is known that unusually severe service is involved. These decisions are reached by working with the vendor and assessing the cost-to-benefit ratio for each problem.

SIZING PROCEDURES

P. S. Buckley has stated that "Satisfactory control usually means sufficiently rapid response to setpoint changes or adequate attenuation of disturbances" (Ref. 3.2). A valve is useless unless it augments the controllability of the process within the scope of the above definition. Control aspects are accommodated in the sizing of the valve, the valve characteristic, and the speed of response as provided by the valve-and-actuator combination.

Proper sizing is the single most important factor in the valve's contribution to control. If it is extreme in size (relative to its task), there will be poor control or no possibility of proper control. Correct sizing involves compliance with design parameters, expected variations and an adequate allowance for contingencies. The engineer must often search beyond the process and mechanical flow sheets for the minimum, normal and maximum operating conditions as well as any special startup, shutdown or emergency situations. Flow rates and the associated pressure drops are the essential ingredients of this survey, but all properties and conditions must be reviewed to assure that all critical sizing variables are accounted for.

Allocation of pressure drop to a control valve in the hydraulic system is a function of the total system dynamic losses and/or energy to be dissipated. Control valve size (and cost) can be minimized with high pressure drop assignments, but pumping costs and equipment ratings are increased dramatically by this technique. Today, with the rising cost of energy, it is imperative to minimize the pressure drop allocated for control purposes. This can be done by selection of high recovery valves (i.e., ball and butterfly types) and by narrowing of the commonly applied allowances for pressure drop. In other words, allow for less safety factor and a more critical design.

The ISA Handbook of Control Valves (Ref. 3.3) summarizes the traditional guidelines for pressure drop allocations. These guidelines are intended for common designs and include proper allowances for contingencies. Deviations from these guidelines should be well thought out to assure operabil-

Table 3-1. Valve Terminology (Ref. 3.3)

Inherent flow characteristic of a valve is defined as the relationship between fractional valve lift and the relative flow through the valve at a constant pressure drop.

Installed characteristic is the actual lift-vs-flow characteristic under system operating conditions where pressure drop is not constant, and is unique to each specific installed system.

$$\text{Rangeability} = \frac{\text{Maximum controllable flow}}{\text{Minimum controllable flow}}$$

$$\text{Turndown} = \frac{\text{Normal maximum flow}}{\text{Minimum controllable flow}}$$

$$\text{Valve Gain } K_v = \frac{\text{Change in flow}}{\text{Change in stem position}}$$

and K_v = the slope of the valve characteristic curve.

$$\text{Valve pressure drop ratio} = \frac{P_V}{P_V + P_L} = P_R$$

where P_V = valve pressure drop;
P_L = system friction losses (piping and other)

ity and control over the expected operating range. Designing for minimum energy consumption, for example, is an area which requires extra attention.

H. D. Baumann (Ref. 3.4) has written on the allocation of pressure drops as low as 5 to 10 percent of the pumped system losses for the newer rotary control valves (which have very good rangeability), and for globe valves with a turndown of less than 5:1. The valve pressure drop may also be reduced, with proper caution and attention, on systems where the dynamic losses are well defined and the load changes minimal. On systems with wide and rapid load changes, it is best to retain higher pressure drops to assure control capability; fortunately, these very dynamic systems are usually the streams with smaller energy content. The significant economies can still be realized in energy savings from the stable high-volume, large-diameter, high-horsepower pumped systems.

Once the designer has established the system operating conditions and limits, the mechanics of C_v calculation are well defined by ISA Standards S39.1 and S39.3, and well explained in Chapter 6 of the *ISA Handbook of Control Valves* (Ref. 3.1) and by the valve manufacturers' sizing literature.

It is desirable to calculate minimum, normal and maximum C_v values. With this data, the normal valve size can be selected and reviewed for rangeability and turndown (see Box), as well as for the valve characteristic, or operational curve.

VALVE CHARACTERISTICS

Most commercially available valves have inherent flow characteristics that fall between the quick-opening and equal-percentage curves as shown in Fig. 3-2. The *inherent flow characteristics* provided by the manufacturers are based on a plot of test results for flow vs. the stem position at constant-

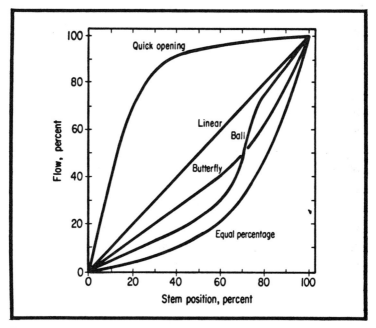

Fig. 3-2. These inherent valve characteristics are commonly applied in the process industries.

pressure-drop operating conditions (per ISA procedures defined in RP-39.2 and 39.4).

The engineer should be acquainted with the term *valve gain*, (K_v), or sensitivity, to appreciate the value of the characteristic curves and their usefulness for control purposes. Gain is the slope of the inherent characteristic curve and relates the change in flow to the change in stem position. The linear valve has a constant gain of 1.0 while the equal-percentage valve exhibits a low gain at the near-closed position that increases to a very high gain at full capacity. The quick-opening characteristic provides the opposite situation with the very high gain at the lower capacity end and essentially zero gain from 50 to 100 percent of stem position. A look at the control loop diagram in Fig. 3-3 will help illustrate the relationship of valve gain to controllability. Ideally, the open loop gain K_L of a stable control loop is constant (Ref. 3.3) and defined as:

$$K_L = K \, K_a \, K_v \, K_p \, K_t = \quad Constant$$

where K is the controller gain; K_a, the valve actuator gain; and K_t, the transmitter gain. All are almost constant over wide ranges of operation. Thus the valve gain K_v and the process gain K_p are the significant variables that must compensate for each other if the control loop is to remain stable.

If these gains are not compensating, it will be necessary to adjust the controller gain (proportional band setting) to correct for the net differences. This becomes impractical when one considers the number of control loops that would have to be adjusted simultaneously whenever changes in throughput or other load variations occur. Thus, good design dictates that the valve must compensate for the changes in process gain over as wide an operating range as possible.

Gain matching through selection of control valve characteristics can be accomplished relatively easily in the control systems that are most often encountered. Consider first the

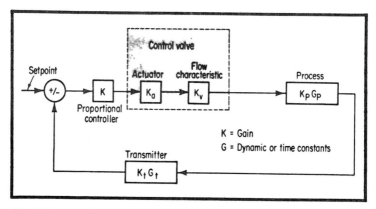

Fig. 3-3. A control loop contains various component gain and dynamic functions. The valve flow characteristic K_v can be critical for good control performance.

case where a control valve is throttling flow between a stable high pressure system and a constant low pressure system with almost no piping or system friction losses other than the control valve itself. This is an example of a loop with constant process gain. A linear characteristic is the logical choice because the valve gain is constant over the entire stem travel.

As a second example, consider a system with piping friction losses relatively higher than the pressure drop across the control valve. This type of process is very responsive at low flows (low system losses) and relatively restricted at the high flows because of the higher system friction losses. The control valve should have a gain that compensates for the decay in process gain, and this characteristic is available in the equal-percentage curve. As a third example, on-off control requires rapid flow changes at the initial valve opening, so the quick-opening characteristic is the proper choice.

Inherent valve characteristics, as supplied by the manufacturer and as determined under laboratory conditions, have been discussesd so far; however a valve usually does not operate in a process under constant pressure drop conditions. One must keep in mind that installed characteristic (the actual inservice relationship between fractional stem position and flow) can be decidedly different from the inherent characteristic.

If a sufficiently high pressure drop was allocated to the valve, then the installed characteristic would be very near to the inherent characteristic of the valve. But, as a lower valve pressure ratio is applied, the inherent characteristic tends to linearize for the equal-percentage valve and to approach the quick-opening curve for a linear valve.

As the installed characteristic curve continues to move to the left in the flow vs. stem position graphs, the characteristic approaches the quick-opening and on-off curves. This leads to a very sensitive and unstable loop that will require frequent controller tuning to accommodate load changes.

Thus, the design must include sufficient pressure drop, say 10 to 30 percent of the system friction loss in a pumped system, to maintain the required installed characteristic and minimize controller adjustments.

In practice, a system can be analyzed for characteristic requirements for the pumped transfer of fluid from one system to another. Figure 3-4 is a natural result of the system hydraulic calculations that are performed early in the design process. The pump supplies the energy to move a fluid against the static head differential, to overcome the friction losses in the piping and to provide the pressure drop necessary for the control valve to perform its function.

The pump head-flow curve is established by the impeller design, and the pump head decays as flow increases. The system friction losses P_s increase with flow rate and are proportional to the square of the flow. The remaining element, static head, is the algebraic sum of the change in elevation head and pressure heads involved in the system. Note that the pressure available for the control valve P_v is represented by the difference between the pump head and the sum of the system friction losses plus the static head.

Data as shown in Fig. 3-4 can be used to develop the installed characteristic curve for the ideal control valve required by the system. The result is a curve similar to the

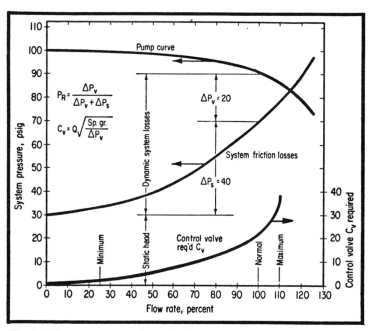

Fig. 3-4. The system curve shows the friction losses for a pumped liquid, and suggests thhat the equal-percentage characteristic would be the best choice whenever the pressure drop across the control valve is small, relative to the friction losses and static head.

inherent equal-percentage characteristic in Fig. 3-2. A valve should duplicate this curve as nearly as practical. In this case, an equal-percentage valve with full-open C_v of approximately 50 would be a reasonable choice for a design flow of 100 percent, a minimum flow of say 25 percent, and a maximum up to 110 percent.

Note that the Fig. 3-4 example comes up with a valve pressure drop of 20 psi, and the system friction loss is 60 psi at the design normal flow ($P_R = 0.33$). This is desirable to assure rapid and certain control. However, if the system is subject to little variation, it may be feasible to dedicate a lesser pressure drop to the valve and conserve energy. As stated above, a minimum of 10 percent of the system loss should be in the valve. This equates to 6 psi, and would permit a pump with a 14-psi-lower head to meet the maximum flow of 100 gpm with a 6 psi control valve pressure drop and a required C_v of

41. Note that the design flow C_v has increased approximately 50 percent because of the reduced pressure drop, and the valve size would have to be increased to obtain the horsepower reduction of approximately 15 percent.

In a practical sense, the information required to fully analyze valve characteristic requirements is often unavailable. And in the interest of economics, application guidelines—rather than extensive analysis—are the basis of engineering practice. In most systems, as in the example discussed in Fig. 3-4, a control valve must be fairly insensitive at the closed positions and responsive at high flow and the equal-percentage trim meets this need. But in cases where the major system pressure losses are in the control valve, the linear characteristic should be selected.

VALVE ACTUATORS

The sizing of a valve and selection of its characteristic relate to the steady-state performance of the control loop. Dynamic considerations must be added to make the loop and the valve compatible with the actual process and the instrument system. Since processes are ever changing and are very much a dynamic function, the control valve must respond in a sufficiently fast manner to influence the process. The control valve actuator and its accessories can furnish these dynamic capabilities. The actuator must also be capable of generating adequate force for valve operation from the control signal and compatible with the environment.

Actuators are available in forms that range widely in cost and performance characteristics. The common spring-diaphragm actuator is the most economical and has the widest use because of its simplicity and proven reliability. Piston actuators are available at a nominal increase in cost and offer advantages in higher force, smaller size, and increased speed. In recent years, the vane type of actuator for rotary type valves has come into prominence and is joining the diaphragm and piston in the work horse team of the actuator lines.

There are also electrohydraulic, hydraulic, and electromechanical actuators that can be applied when their capabilities are required—and when the added cost is justified; these represent less than 1 percent of the applications and therefore will not be dicussed further.

The common actuators (spring diaphragm, piston, and vane types) are all operated with a clean air supply as the actuating medium. To date, a competitive electrically-controlled actuator has not become available, so it appears that the cost and complexity of the electric actuator will preclude its wide use for a number of years.

A control signal range of 3–15 psig is usually selected to operate control valves. When the instrument system is electronic, a transducer is required to convert the 4–20 mAdc (or other) output signal to a 3–15 psig pneumatic signal for input to the valve. Either electronic or pneumatic controls can thus be applied to the same actuator, but there is a problem in the I/P conversion.

The nominal 5:1 signal range with the live zero (3–15 psig) has historically been put into actuators in a manner that utilized the 0-3 psig and 15-18 psig fringes of the controller output signal for excess stem loading to assure tight shutoff at the seat ring. This has worked well with no impairment of control; however, with electronic controls and more so with digital techniques, the control signal may be an absolute 4 – 20 mAdc or other standard signal without overrange capability.

A transducer with the 4–20 mAdc input, when calibrated normally, will respond with a 3–15 psig output signal. The output will not fall below 3 psig output signal. The output will not fall below 3 psig nor rise above 15 psig to reproduce the overrange condition available from pneumatic control systems. If this output overrange is not available, the problem must be acknowledged when valve actuators are specified. Tight shut-off commonly obtained with the benefit of a 3-pound excess loading must be obtained by resilient seats, closer metal tolerances, initial shutoff at 6 or 12 psig, or some other technique.

Environmental effects on the actuator vary widely from one industry to another. Petrochemical plants usually benefit from the use of PVC coated tubing, and good paint maintenance. More and more actuators are being fabricated from aluminum and this is becoming a significant valve selection problem. In many plants, aluminum actuator housings are undesirable because of exposure to caustic soda and other chemicals that cause rapid corrosion.

DYNAMIC RESPONSE

Actuator speed of response is determined by the diameter and the length of the pneumatic signal tubing, and by the volume of the diaphragm housing. Valves over three inches in body size with spring diaphragm actuators generally exhibit a slow response; consequently, a volume booster or valve positioner must be installed to provide adequate response. This limitation holds true only for diaphragm actuators; most other actuator types have relatively small volumes and/or utilize integral valve positioners.

A volume booster is a pneumatic relay that has a small volume at the inlet side and a high capacity air regulator to provide a high volume output signal. It is often a 1:1 repeater—but can be 1:2 or 1:3. It overcomes the lag caused by a long length of tubing in combination with the large volume of a diaphragm actuator, Fig. 3-5.

The valve positioner is a valve-mounted accessory that drives the control valve plug to the precise position requested by the input control signal. It serves to overcome packing friction, and to improve speed of response.

While the list price of a positioner is several times that of the booster, the installed costs are not much different. The positioner offes extra benefits in the form of predictability, improved sensitivity, signal reversal option, or split ranging

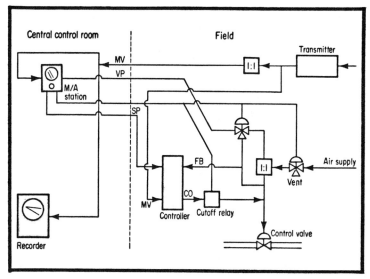

Fig. 3-5. Pneumatic transmission lags can be minimized by pressure repeaters or volume boosters (marked 1:1) on the valve and/or transmitter signals. In the four-pipe system shown here, a field-mounted local controller also helps to minimize lag.

for the application of two valves on a common control signal. Positioners are usually specified, rather than boosters, because they offer the greatest flexibility.

Specification policy on positioners varies among users from installation on every valve to only on those larger than a certain size, or only when required for specific control problems (e.g., split range, high resolution, special characterization, high packing loads). There have been cases where positioners caused instability in fast flow control loops. The positioner—like the booster—is a capital cost item and an *air consumer*; therefore, it should not be specified unless needed.

While discussing valve positioners, it is useful to mention an accessory known as the bypass switch. All too often, this accessory is specified when the bypass can never be used. Effectively, the bypass switch removes the valve positioner from the signal path between the controller and the valve actuator. This is sometimes desirable for minor maintenance and to investigate loop disturbances.

277

If the positioner is truly applied as a 3–15 psi input and output device, then the bypass valve may be used; in any other case, use of the bypass is not acceptable. Consider a bypass on a reverse acting positioner: The valve will respond with movement in the wrong direction! Where the positioner provides signal amplification or split ranging, the switch cannot be used without complete sacrifice of control. Be cautious about the availability and use of this bypass switch.

It appears that the majority of new control valve designs may require valve positioners because the actuator springs will no longer match the conventional 3–15 psig controller signal, and it is necessary to utilize the positioner as a pressure multiplier. This is a result of an effort to standardize on a smaller number of actuators to serve a wider size range of control valves. Manufacturers and users have a common interest in this effort. Manufacturing ease and reduction in required spare parts will result in a lower unit cost for the user. Almost everyone benefits from this trend—except possibly the mechanic who now has a more complex calibration task.

FAILSAFE ACTIONS

During the initial design of a process unit, it is usual practice to analyze each valve for failure position in the event its power source is lost or interrupted. This decision should be viewed from two viewpoints: What is the desired action in the event of a plant-wide failure? What is the desired action if only the valve in question loses its signal source?

The choices for action upon failure are:

- Fail open.
- Fail closed.
- Lock up in the last position.

The first two failure actions are best accomplished through springs that drive the valve to the desired position when the opposing power source is lost. Alternately, a system of air reservoirs with valves and relays can load the proper side of a piston actuator. Lockup is accomplished in a similar manner or with a simple pneumatic relay.

The spring technique is the most reliable. The reservoir/valves/relay systems offer more opportunities for failure, at a time when proper operation is essential. Since reliability is a primary consideration, a simple design is far superior.

However accomplished, control valve selection must be carried out in a manner compatible with the overall project design schedule. Methods vary appreciably among the numerous engineers and firms in the field. From the user's viewpoint, a complete design will encompass the process, mechanical and control aspects and accommodate each in a satisfactory manner.

Valve selection is greatly influenced by the application engineer's personal experience and proven guidelines and recommended wherever possible. A lack of time and information during the design period often precludes the theoretical approach and does emphasize practical knowledge and its application. However, for the novice and for those seeking additional background, the *ISA Handbook on Control Valves* provides a complete reference.

REFERENCES

3.1 Driskell. L. R., "Sizing Control Valves," *ISA Handbook of Control Valves*, 2nd ed., ISA, Pittsburgh, PA, 1976, pp. 180-205.

3.2 Buckley, P. S., "Design of Pneumatic Flow Controls," *Proceedings of the 31st Annual Symposium on Instrumentation for the Process Industries*, Texas A & M Univ., 1976, pp. 61-73.

3.3 Moore, R. L., "Flow Characteristics of Valves," *ISA Handbook of Control Valves*, 2nd ed., ISA, Pittsburgh, PA, 1976, pp. 165-179.

3.4 Bauman, H. D., "How to Assign Pressue Drop Across Control Valves for Liquid Pumping Services," *Proceedings of the 29th Symposium on Instrumentation for the Process Industries*, Texas A & M Univ., 1974, pp. 46-53.

3.5 Buckley, P. S., "A Control Engineer Looks at Control Valves," *Proceedings of the 1st ISA Final Control Elements Symposium*, Wilmington, Del, 1970.

DUANE G. WOLTER is a senior engineer in the Process Control group of Shell Oil Co., Houston, TX. Article is based on a paper presented at the ISA/77 Spring Conference, Anaheim.

Control Valve Dynamics

M. ADAMS

Operation of a valve and actuator is influenced by a number of factors in the valve, actuator and process under control. These various influences must be taken into account when a valve and actuator system is specified to insure that it will work effectively when installed.

Spring and diaphragm actuators are used widely in the process control industry because of their low cost, flexibility, simplicity and reliability. In many cases, a properly configured spring and diaphragm actuator can perform excellent valve throttling control without requiring positioners or boosters.

To select a diaphragm actuator properly for proportional control, it is necessary to know the characteristics of the controlling instrument, control system, valve body and the actuator itself, since each component exerts a separate and substantial influence on loop performance. This article examines these influences in detail, using a case study of a globe valve with an air-to-open/spring-closed actuator, Fig. 3-6, as an example.

INSTRUMENT INFLUENCE

A controlling instrument has the task of providing air pressure to the actuator and controlling the amount of actuator stroke. The most important characteristics of such a controller are its output span and volumetric output capacity.

Output span determines the range of supply pressures available to the actuator. A normal proportional pneumatic controller has either a 3–15 or 6–30 psi signal, limiting the actuator to a 12 or 24 psi operating span. The actuator must move the valve from the shutoff position to full open within that operating span.

If reset action is added to the controller, it gains the capability of spanning from almost zero psi to nearly its full supply pressure (usually 20 or 35 psi). Pneumatic controllers generally have pneumatic relays, sized large enough to maintain normal actuator stroking speeds.

Electronic controllers normally are interfaced to control valves by means of electropneumatic (I/P) transducers. Extreme care must be taken when applying I/P transducers directly to diaphragm actuators, because the transducers almost never span beyond their 12 or 24 psi range; also, many I/P transducers are not designed to output the volume of air necessary to stroke an actuator with adequate speed.

Bench set is generally defined as the stroking range of the actuator neglecting external influences. An actuator with a 6–15 psi bench set would begin to stroke at 6 psi and be at full travel at 15 psi casing pressure. Because all external influences are neglected, the bench set is only a calibrating and testing procedure which insures that the actuator will stroke properly when installed.

FORCES IN OPPOSITION

The actuator is a source of several influences on system parameters; the most obvious of these are forces exerted by the spring and diaphragm, which provide the loading force for the valve plug.

A diaphragm is a sheet of cloth-reinforced rubber against which air pressure works to create a force. The magnitude of

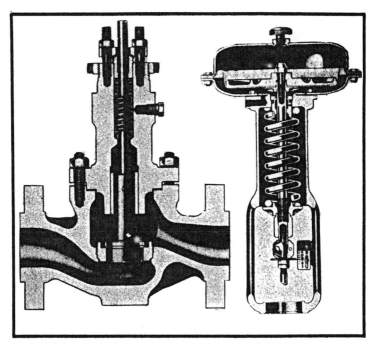

Fig. 3-6. This 2-in. single-port unbalanced globe valve (left) and an air-to-open, spring-closed actuator (right) will be used as a case study of valve dynamics.

this force is the product of the diaphragm's effective area and the pressure differential across it. In most cases, only one side of the diaphragm is pressurized; the other side is exposed to the atmosphere. Therefore, diaphragm force F_d can be calculated:

$$F_d = P_L A_d \qquad (3.1)$$

Where P_L is the casing loading pressure and A_d is the effective diaphragm area. Other definitions are given in the nomenclature.

Diaphragms used in most actuators are molded so as to maintain an almost constant effective area throughout their travel. Simple flat-sheet diaphragms, however, are subject to considerable area changes during travel which must be considered in force calculations.

The actuator spring opposes diaphragm motion. The spring usually is compressed when the actuator is at its zero position, thereby delaying actuator travel until a predetermined pressure has been applied to the diaphragm. When loading pressure on the diaphragm increases, the spring compresses as the actuator diaphragm goes upward. When diaphragm pressure is removed, the spring returns the diaphragm to its initial position.

Hooke's Law defines the force change required to compress the spring. Simply stated, this force change is the product of the spring rate and the change in displacement on the spring:

$$\Delta F_s = K_s \Delta t \qquad (3.2)$$

To calculate the change in air pressure necessary to compress the spring:

$$\Delta P_L = \frac{\Delta F_s}{A_d} \quad \Delta P_L = \frac{K_s \Delta t}{A_d}$$

A valve body also requires and creates significant forces. In order to effect shutoff, the plug must be forced tightly against the seat. The required force is known as seat load. Seat load requirements vary from valve to valve, depending on design, seat style and application. For an air-to-open valve, seat load is derived from precompression of the spring.

Friction forces also are present in the valve body. These forces come from packing, seals and guiding surfaces. The weight of the valve plug and stem also should be considered for larger valve body assemblies.

UPSTREAM AND DOWNSTREAM

The main process influence is the static unbalance created at valve shutoff. This unbalance is due to pressure differences across the plug at shutoff, Fig. 3-7.

In flow-open constructions, the upstream pressure registers under the plug creating an upward force. The magnitude

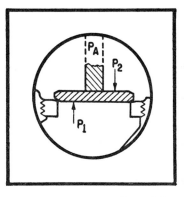

Fig. 3-7. In the shutoff position, upstream pressure P_1 pushes against the valve plug, trying to open the valve, while down-stream pressure P_2 acts in the opposite direction. The difference between the forces is called the net static unbalance force; in the case study, the unbalance force is 628 lb.

of this force is the product of the upstream pressure P_1 and the orifice area A. Downstream pressure P_2 acts on the upper side of the plug creating a downward force. The downward force is equal to the product of P_2 and the orifice area minus stream area A_s. The result is the net static unbalance force, which can be expressed:

$$F_{us} = [P_1 A_0] - [P_2 (A_0 - A_s]$$

or

$$F_{us} = A_0 \Delta P + P_2 A_s \tag{3.4}$$

Static unbalance is by no means the only process influence on the valve. In all open positions of the valve, dynamic

Table 3-2. Nomenclature

A_d	effective diaphragm area, in.²
A_s	valve stem area, in.²
A_o	valve orifice area, in.²
C_v	liquid flow coefficient
F_d	force of diaphragm, lb
F_s	force of spring, lb
F_{us}	static unbalance, lb
F_{ud}	dynamic unbalance, lb
K_s	mechanical spring rate, lb/in.
NSF	normalized stem force, dimensionless
P_A	atmospheric pressure
P_L	casing loading pressure, psi
P_I	upstream or valve inlet pressure, psi
P_2	downstream or valve outlet pressure, psi
ΔP	valve pressure differential (P_I-P_2), psi
t	valve travel, in.

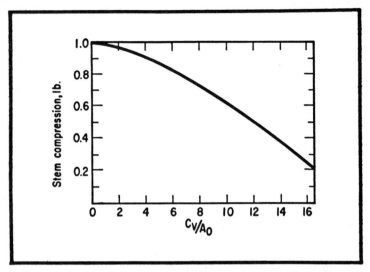

Fig. 3-8. Fluid moving past the valve plug causes a pressure differential across the valve; the resulting force always acts against the valve stem. This figure shows the force characteristic for a 2-in. globe valve with a constant pressure differential at all openings.

forces act on the plug because of fluid momentum. The forces vary in magnitude and direction, depending on valve style. In all cases, however, flow forces are proportional to the pressure differential across the valve. A typical single-port flow-up unbalanced globe valve exhibits an inherent force characteristic, as shown in Fig. 3-8. The magnitude of the dynamic force decreases as the valve travels, but maintains the same direction, i.e., a compressive load on the stem.

Figure 3-8 represents the force characteristic for a valve with a *constant* flowing pressure differential at all openings. This characteristic is not found in most applications, however, and corrections must be made for pressure drop variations vs plug travel. To evaluate dynamic stem force in a specific case, it is necessary to determine the flowing pressure drop, valve flow coefficient C_v, port area, stem area and the downstream pressure.

The normalized stem force *NSF* can be read on the vertical axis of Fig. 3-8 for specific values of the ratio C_v/A_o; the actual stem force is then found by solving the following equation for F_{ud}:

$$NSF = \frac{F_{ud} - A_s P_2}{A_o \Delta P}$$

$$F_{ud} = (NSF \times A_o \Delta P) + A_s P_2 \qquad (3.5)$$

Repeating this procedure for a number of points in the valve travel yields a number of F_{ud} values which can then be plotted to show the installed dynamic unbalance characteristic of a particular single-port globe valve.

CASE STUDY

To demonstrate the concepts presented so far, an actual case study will be explored. Figure 3-9 illustrates a 2-in. single-port unbalanced globe valve in pressure reducing service. The valve has a 2-in. port and a ½-in. stem.

Downstream loads require varying flows at a constant pressure of 35 psi. The valve must shut off against 200 psi, and flowing pressure drops across the valve vary from 35 to 165 psi. Pump head, valve ΔP and system loss curves are shown in Fig. 3-10.

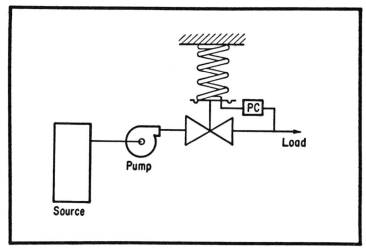

Fig. 3-9. Block diagram of the case study situation. The 2-in. globe valve is in pressure reducing service and controlled by an actuator with a 6-30 psi signal.

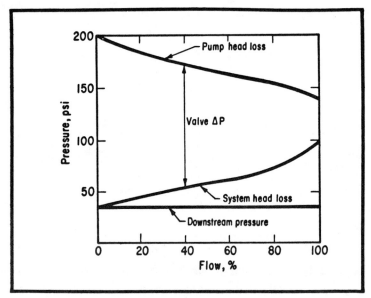

Fig. 3-10. Loss curves for the case study valve; output or downstream pressure is maintained at 35 psi in this application.

The actuator controls the valve with a 6–30 psi signal; the actuator diaphragm has 105 square in. of constant effective area.

Using actual data from the case study, a spring will be selected for the actuator, a bench set determination performed and the effects of process influences examined. Finally, the performance of the valve in service will be analyzed.

SPRING SELECTION

Calculating valve plug forces at the extremes of valve travel—full open to full closed—is the first step in spring selection.

Static unbalance in the closed position is calculated according to Equation 3.4:

$$F_{us} = A_o \, \Delta P + P_2 \, A_s = 628 \text{ lb}$$

During operation, the actuator must keep the valve closed until 6 psi is applied to the diaphragm area. Six psi, acting on a 105 square in. diaphragm area, creates a force of

630 lb that tries to open the valve. At the same time, the static valve plug unbalance of 628 lb also is trying to open the valve. The total of these forces (1,258 lb) must be mechanically loaded into the spring to maintain the valve in the closed position at a pressure of 6 psi, Fig. 3-11.

Dynamic force in the open position is calculated from Equation 3-10 and Fig. 3-8. Since the C_v/A_o ratio is 15.9, the *NSF* from Fig. 3-8 is 0.2. Solving Equation 3.5 for dynamic force yields:

$$F_{ud} = (NSF \times A_o\ \Delta P) + A_s\ P_s = 35 \text{ lb}$$

This force is in the upward direction and tends to open the valve.

In the full-open position, the diaphragm exerts 3,150 lb (105 sq. in. × 30 psi). To maintain force balance, Fig. 3-12, total compression in the spring at full open must equal the sum of these two forces (3,150 + 35), or 3,185 lb.

For satisfactory performance in a 6 to 30 psi operating range, actuator spring compressions must be 1,258 lb at zero travel and 3,185 lb at full travel of 1.125 in. The ideal spring rate can be calculated by solving for K_s in Equation 3.2:

$$\Delta F_s = K_s \Delta t$$
$$1{,}927 \text{ lb} = K_s \times 1.125 \text{ in.}$$
$$K_s = 1{,}712 \text{ lb/in.}$$

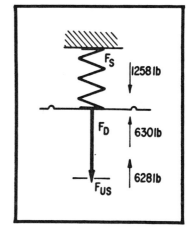

Fig. 3-11. Force-balance diagrams illustrate the direction and magnitude of forces acting against each other. In the shutoff position, when the actuator is at its initial position, the spring must be preloaded to 1,258 lb to keep the valve shut against a 628-lb static unbalance force and 630 lb from the actuator diaphragm at 6 psi.

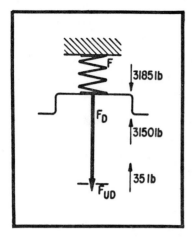

Fig. 3-12. At full valve travel, the spring must have a 3,185-lb load to equalize the dynamic unbalance of 35 lb caused by pressure differential across the valve and the 3,150-lb diaphragm force.

If a spring with the exact rate required is not available, a spring should be selected that has a rate as close as possible to (but not exceeding) the ideal. In this case, a spring with a rate of 1,670 lb/in. was chosen.

BENCH SET

During a bench set test, the operating range of the actuator is measured without the external influence of the valve and process. In the zero position, Fig. 3-13, the actuator must move against the 1,258-lb spring preload without the help of the 628-lb valve static unbalance force. In order to move the diaphragm, a loading pressure of 12 psi is required.

To completely stroke the actuator and totally compress the spring 1.125 in. to 3,136 lb—1,258 lb preload plus 1.125

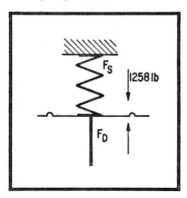

Fig. 3-13. In a bench set, external influences are ignored; the force diagram for the shutoff position, therefore, shows only the spring and diaphragm forces.

in. at 1,670 lb/in. —the actuator must develop 29.8 psi loading pressure, Fig. 3-14. The range of the spring stroking pressure, therefore, is 12–30 psi, and this is defined as the bench set.

INSTALLED CHARACTERISTICS

Forces at various increments of actuator travel can be determined by expanding the case study calculations. For example, process characteristics at 10 percent increments of total flow are shown in Table 3-3; these data were calculated using the techniques described previously.

To have equilibrium in every valve position, the downward force of the spring must be always equal to the sum of the two upward forces—diaphragm pressure and dynamic plug force. The dynamic plug force is calculated with Equation 3-5; to determine actuator loading pressure, divide the diaphragm pressure requirements by the effective diaphragm area.

Figure 3-15, derived from such calculations, illustrates the relationship between stem force and valve travel. Comparing Fig. 3-15 to Fig. 3-8, which shows the inherent force characteristic, major changes can be noted in force magnitude, especially at the wide-open position. These changes are due to the effects of decaying pressure drops across the valve as it opens.

The real proof of the actuator spring selection and bench set techniques is shown in Fig. 3-16, which plots actuator

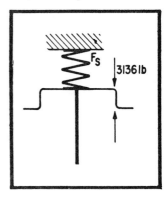

Fig. 3-14. Bench set force diagram at full travel also shows only the spring and diaphragm forces.

Table 3-3. Table System Parameters

Total flow	10%	20%	30%	40%	50%	60%	70%	80%	90%	100%
Valve inlet pressure P_1 (psi)	193	187	182	178	174	170	165	159	150	140
Flow coefficient C_v	2.4	4.9	7.7	10.7	13.9	17.3	21.2	26.1	33.8	50.0
Valve differential pressure ΔP (psid)	153	142	132	122	114	105	95	82	62	35
Dynamic plug force F_{ud} (lb)	482	434	400	359	324	283	244	190	121	35
Spring force F_s (lb)	1,483	1,859	2,143	2,277	2,410	2,477	2,561	2,644	2,744	3,137
Diaphragm force F_d (lb)	1,000	1,425	1,743	1,918	2,086	2,193	2,317	2,454	2,623	3,102
Loading pressure P_L (psi)	9.53	13.57	16.60	18.27	19.87	20.89	22.07	23.37	24.98	29.54
Valve travel t (in.)	0.14	0.36	0.53	0.61	0.69	0.73	0.78	0.83	0.89	1.125

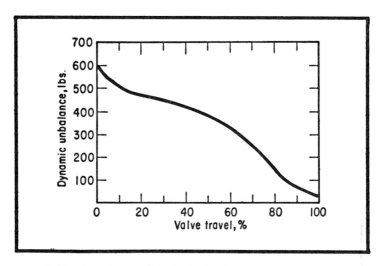

Fig. 3-15. Plug force characteristics for the installed valve are quite a bit different than those shown in Fig. 3-8. The force against the valve stem drops off at the wide-open position because the pressure differential decreases with the case study valve.

signal range, bench set range and actual operating range for the installed valve. In this figure, installed valve travel is plotted as a function of diaphragm loading pressure. As predicted, the actuator begins to stroke at 6 psi and reaches full travel at 30 psi.

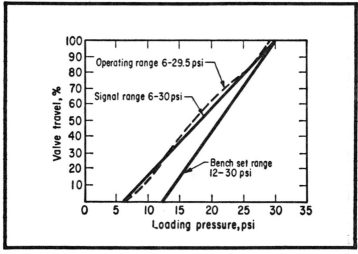

Fig. 3-16. The valve operates perfectly in service: it begins to open at a loading pressure of 6 psi and is wide open at 30 psi.

293

MINOR DETAILS

Several factors which would have complicated the procedure were omitted from the case study. These include friction from bearings and the weight of moving parts.

These factors can be considered by simply adding them into the force balance diagrams, noting their magnitude and direction. Frictional forces always work in opposition to the prime mover, and frictional forces should be considered if packing other than TFE is used—especially laminated graphite or semimetallic packings. Piston rings used in balanced cage-guided valve bodies also contribute significant frictional forces.

The weight of moving actuator parts generally is negligible and can be ignored; the weight of valve plugs, especially those in larger valves, should be considered.

If all influences and considerations are evaluated and the spring selection is carefully made, a spring and diaphragm actuator can be configured to serve as an extremely good proportional control element without the need for positioners or boosters.

MARK ADAMS is a product analyst for Fisher Controls Co., Marshalltown, IA. Article is based on a paper presented at the ISA/77 Spring Conference, Anaheim.

Characterized Valve Actuators

Q.V. KOECHER

Traditionally, control valve manufacturers have designed their products with characterized valve plugs driven by linear, no-gain actuators. This practice requires a compromise between flow characterization and other valve body design parameters to the detriment of performance. By characterizing the actuator, more desirable features of valve construction can be provided while permitting greater latitude in characterization. The result often can be a reduction in cost with improved durability and performance.

Flow or pressure control of any fluid process requires three separate functions: measurement, computation and correction. These functions are usually performed by a process transmitter, a controller and a control valve, respectively. Both the transmitter and the controller are generally linear devices; i.e., the ratio of input to output is constant. The control valve, however, generally is characterized to provide the best stability over its control range. Its ratio of input to output may be characterized linear for level control, characterized equal percentage for flow or pressure control, or it may be characterized with a parabolic or some other function to fit a particular requirement.

Traditionally, characterization has been provided in the design of the valve body and plug. The actuator mechanism, the device which converts the control signal into mechanical motion, is almost always a linear, no-gain (1:1 ratio) device. One of the more common designs, Fig. 3-17, is a plug valve contoured to provide the desired flow characteristic. The actuator is driven by a piston or diaphragm that provides equal increments of stroke for equal increments of signal change. In a characterized ball valve, Fig. 3-17B, the orifice in the ball is contoured to achieve the required flow characteristic. Again, rotation of the ball is accomplished by an actuator mechanism which provides essentially equal increments of rotation for equal increments of signal change.

The design or selection of any control valve must consider the parameters of physical stress (expansion, contraction, bending, vibration and line pressure), rangeability (the maximum and minimum process conditions which must be controlled), maintainability (parts replacement, ease of service frequency of service or inspection) and cost (both initial and continuing). Each parameter must be compromised with others. For the valves in Fig. 3-17, flow characterization also must be compromised with rangeability, physical stress and cost.

By providing characterization in the actuator, the latitude for design or selection of the valve body is increased with a consequent improvement in overall performance. This improvement often can result from improved controllability, rangeability, durability, maintainability, cost, or a combination of these factors.

CHARACTERIZING ACTUATORS

The character of a control valve is measured by its flow characteristic—the flow rate produced at equal increments of position change throughout the range of the valve. This flow rate is most often expressed as a flow coefficient C_v, defined as the amount of water in gallons per minute at a specific

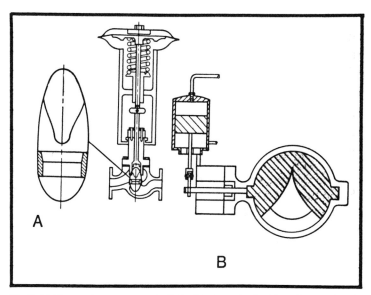

Fig. 3-17. The most common method of achieving a desired flow characteristic is to contour the valve plug to the desired flow function. A) In a characterized plug valve, the plug is shaped to a particular contour. B) In a characterized ball valve, the orifice in the ball is contoured.

gravity of 1.0 and a temperature of 18 °C, which will produce a pressure drop of 1 psig across the valve. Every valve has a finite flow coefficient for each valve position. The flow characteristic, therefore, may be illustrated as a curve which will be either linear, equal percentage or some other shape depending upon the function of the valve.

Figure 3-18 illustrates a typical gain curve of a characterized valve. It is obvious that optimum control can be obtained only at one point on this curve. Process changes which require the valve to move to the right of this point on the curve would tend to result in instability; changes requiring the valve to move to the left on the curve would result in slower response and progressively more sluggish control. By limiting the gain (flattening out the curve), better controllability can be obtained throughout the range of the valve.

In a characterized actuator, the flow characteristic is referenced to the control signal rather than valve position. A very real advantage of a characterized actuator is its ability to

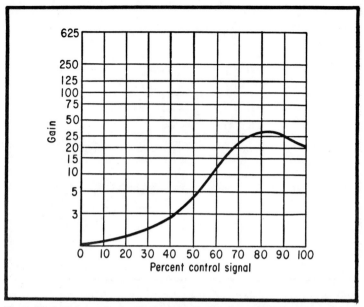

Fig. 3-18. A characterized valve may have nonuniform gain, as shown here. Optimum control can be obtained at only one point on this curve. Changes which shift the valve to the right tend to result in instability while changes to the left detract from performance.

modify the flow characteristic of a valve to provide a more uniform gain throughout its range.

Two different types of characterized actuators have been successfully applied using this concept. Figure 3-19A illustrates a variable ratio actuator (Ref. 3.6) designed to modify the flow characteristic and gain of the valve with which it is used. This is accomplished through a mechanical arrangement of linkages. Figure 3-19B illustrates a speed control toggle-type actuator (Ref. 3.7) that characterizes flow with reference to increments of time (duration of signal) as well as signal amplitude.

A CASE HISTORY

A variable ratio actuator control valve was first used for pressure control in a liquids pipeline pump station. In this application, the control valve is part of a control system that monitors process parameters. The control valve is used to throttle only when one of these parameters exceeds its set limits. Normally, the valve is wide open or "off control." In this position, any pressure drop across the valve is wasted energy. Consequently, one design parameter called for a valve with a minimum full-open pressure drop.

An additional requirement was the capability to throttle out process upsets of at least one pump head at minimum flow. Thus, the consideration of rangeability was added to the design requirements. Ignoring the flow characteristic requirement, the valve body was selected on the basis of these two parameters. A full conduit ball valve was the obvious choice

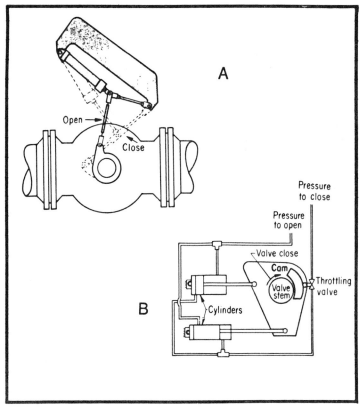

Fig. 3-19. Two types of characterized actuators have been developed which can modify the flow and gain characteristics of a valve. A) This variable ratio actuator uses mechanical linkages to modify the flow characteristic as a function of actuator position. B) A speed control toggle-type actuator makes the flow characteristic a function of time and signal amplitude.

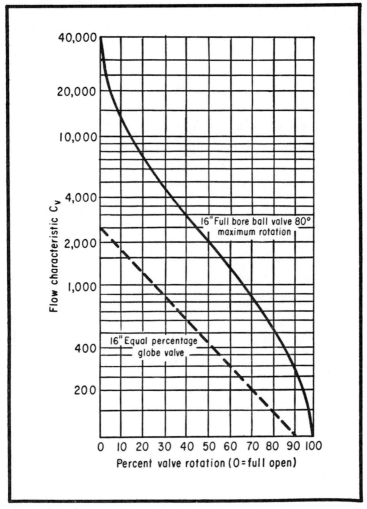

Fig. 3-20. A 16-in. full conduit ball valve was chosen for the pipeline application because it had a minimum pressure drop when full open and could throttle out process upsets of at least one pump head at minimum flow. Flow characteristics of this valve and an equal percentage globe valve are shown.

from a consideration of minimum full-open drop and, with less than full rotation, it met the requirements for maximum throttling.

A 16-in. valve was chosen for this first application. The design and concept, however, is now in successful use with full conduit ball valves of 4-in. through 30-in. sizes.

The maximum flow characteristic, Fig. 3-20, for the 16-in. ball valve is approximately 44,000 which, with line reducers and at design flow rate, would produce less than 0.5 psig drop across the valve. Minimum flow characteristic is less than 100, resulting in a tremendous rangeability for the valve.

Figure 3-20 also shows the flow characteristic of a typical equal percentage 16-in. globe-type valve. While it was desirable to emulate the flow characteristic of the globe valve, the application required the temendous rangeability and low full-open drop of the full conduit ball valve.

The ball valve flow characteristic curve reveals that it is already an equal percentage valve from about 20-deg rotation through about 6-deg rotation. To get the valve quickly into this control range, the first consideration in actuator design was to provide an extremely high ratio of signal-to-rotary motion between 0 deg and 20 deg. This would permit quickly bringing the valve into control with the first increments of signal. The result of this high signal-to-rotation ratio may be seen in Fig. 3-21: A flow coefficient of 7,000 which brings the valve into the equal percentage control range, is achieved with the first 10 percent of signal.

The ball valve gain increases rapidly as it moves toward its closed position. The second actuator design consideration, therefore, was to reduce actuator gain as the valve approached the closed position. By providing a relatively narrow change in gain throughout the signal span, extremely high controller gain could be utilized without the danger of instability at any control point.

To take advantage of the equal percentage flow characteristic of the valve, it was desirable to maintain a relatively constant actuator gain through the mid-range of valve positions. The resultant signal to process gain characteristic curve, Fig. 3-22, shows that this gain curve peaks at about 80 percent of signal. Remembering that gain is proportional to the square of the flow rate, it is apparent that this peak will shift

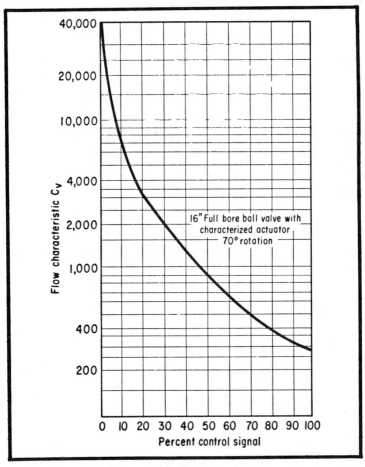

Fig. 3-21. When a variable ratio actuator was installed, the flow characteristic of the 16-in. ball valve shown in Fig. 3-20 became much more responsive. It reached a C_v of 7,000 twice as fast as the normal valve.

with change in flow. Nevertheless, the control loop can be tuned at whatever peak gain position corresponds to the chosen flow rate. Controller gain, therefore, may be adjusted solely with reference to process dynamics without any danger of instability due to overcontrol of the valve.

PROVEN PERFORMANCE

By utilizing the design concept of a characterized actuator, many of the advantages outlined earlier were realized.

Also, by restricting the number of design parameters the valve would have to meet, a valve could be selected that was physically durable under all the stresses of the application, including occasional cavitation. This choice resulted in a valve which—because of simple construction—required minimum maintenance and had a relatively low original cost.

Designing the actuator to utilize much higher controller gains resulted in additional and unanticipated advantages. Much higher valve speeds could be used without the danger of instability, and these high valve speeds made possible the overtaking and controlling of upset conditions heretofore not possible in liquids pipeline process control.

Tests performed at the Katy Station, Rancho Pipe Line, March 21 and March 22, 1975, show the improved valve performance, Fig. 3-23. At this station, maximum throttling could be achieved within 70 deg of valve rotation; therefore, valve rotation was limited to 70 deg. By spreading the signal span over 70 deg instead of full-valve rotation, an improve-

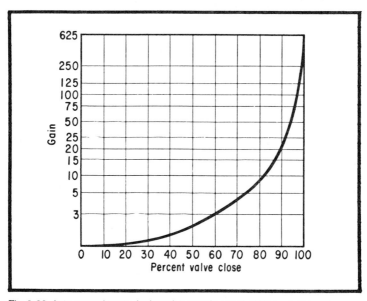

Fig. 3-22. Actuator gain was designed to remain constant through the mid-range and to reduce as the valve approached its closed position. This permitted extremely high controller gain without the danger of instability at the control point.

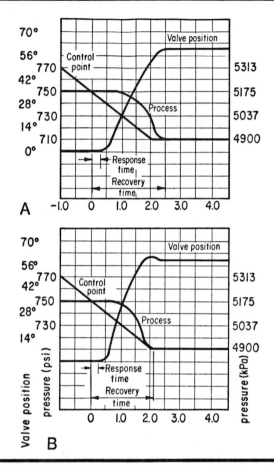

Fig. 3-23. Over 60 tests were performed on station discharge control to simulate a variety of process upsets. The simulated upsets were designed to make the characterized actuator and ball valve perform under "worst case" conditions; i.e., through the full range of valve travel and at maximum throttling. Regardless of the severity of process upsets or controller gain settings, recovery time was approximately two seconds in every case. Four representative tests are shown here, and test parameters were as follows: (see next page)

ment in sensitivity was achieved. The valve actuator was hydraulically driven and capable of moving the valve 70 deg in 3 seconds.

It was found that controlling on the proportional function of the controller with a minimum of reset provided the best performance.

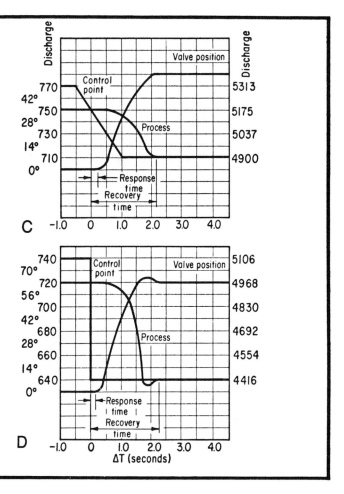

A. controller gain—20; simulated process upset ramp—20 psi/s; control point change—from 20 psig above to 40 psig below process.
B. gain—50; ramp—20 psi/s; control—20 psig above to 40 psig below.
C. gain—20; ramp—40 psi/s; control—20 psig above to 40 psig below.
D. gain—20; ramp—100 psi/s step change; control—20 psig above to 80 psig below.

All tests were stated with the valve full open or "off control"; i.e., the process was within the control parameter. The control point was then ramped at a predetermined rate to a point which required the valve to move through most of its travel to bring the process back into control. It was felt that these conditions reflected "worst case" for this application. In addition, the control valve would be required to demonstrate its controllability near the maximum throttling position, another "worst case" condition. All tests, except Fig. 3-24, were made on station discharge control; the test shown in Fig. 3-24 was made on station suction control.

Figure 3-23A illustrates the results of a simulated process upset of 20 psi per second with a controller gain of 20. Loop response time was about 0.4 second, and the valve brought the process back into control without overshoot in 2.5 seconds.

Figure 3-23B illustrates the same test with a controller gain of 50. Loop response time was slightly less, and the valve was able to recover the process in 2.0 seconds. Again there was no process overshoot. However, a change in valve position after recovery indicates the process had not yet stabilized. It is of interest to note that the control point and process coincide at the 2.0 second mark. It may be concluded, therefore, that the control valve can overtake process upsets, and thus limit the amount of deviation, at rates up to 20 psi per second.

In the test shown in Fig. 3-23C, the gain was set back to 20 and the simulated process upset was ramped at 40 psi per second—a rate probably in excess of that likely to be experienced in actual operation. The valve, however, recovered smoothly in about 2.0 seconds.

Using the same gain of 20, a 100 psi step change was made in the control pont, Fig. 3-23D. Recovery time for this extreme upset was still less than 2.5 seconds.

Since recovery times of all the tests illustrated (plus some 60 additional tests made) indicate a recovery time of approximately 2 seconds regardless of the rate of process change, it might be concluded that this is the dynamic recovery time of the process. The conclusion supports earlier statements made relative to loop tuning at peak gain.

Figure 3-24 illustrates the control valve's ability to overtake most anticipated process deviations and limit the amount

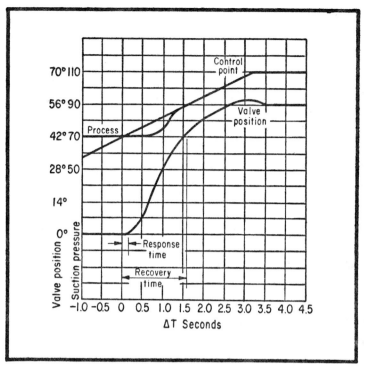

Fig. 3-24. This test was performed on station suction control. With a gain of 20, the valve was able to overtake a ramp change of 12 psi/s from 20 psig below to 40 psig above the control point.

of this deviation. This test was performed on station suction control. The control valve was able to overtake the process deviation of 12 psi per second and affect recovery in about 1.6 seconds. Maximum deviation was limited, in this case, to about 8 psig. Again, the valve was required to make some adjustment to the process as it stabilized near the 3.0 second mark.

OBSERVATIONS

The performance improvement here is apparent. Process recovery time was optimized to the limit imposed by the process dynamics. Controllability at all valve positions was considerably improved. The valve achieved the low drop "off control" objective as well as the extreme rangeability desired.

By eliminating flow characterization from consideration, a better choice of valve body was made; by characterizing the actuator to provide the required flow characteristic with the selected valve body, improved performance was obtained.

But this does not suggest that a characterized actuator is the best design approach to all control valve applications. It should be considered when characterizing the valve body requires serious compromise with other design factors. Where rangeability, structural design of the valve body, maintainability or cost are major factors, the design or application engineer may well find that selecting a valve body for those factors and then characterizing the actuator to achieve the desired flow characteristic will prove advantageous.

We wish to thank J. A. Scheineman and Shell Pipe Line Corp. for providing facilities and conducting the field tests of the variable ratio actuator; and T. M. Fontaine, E. D. Zientara, D. Marcus and Taylor Instrument Co. for providing instrumentation and compiling test data.

REFERENCES

3.6 The variable ratio actuator is the invention of J. A. Scheineman, Shell Pipe Line Corp. Patent is held by Shell Pipe Line.

3.7 The variable speed actuator is the invention of Marshall Bagwell and Ray Phillips, Colonial Pipeline Co. Patent is held by Colonial Pipeline.

QUENTIN V. KOECHER is President of Custom Controls Co., Bellaire, TX. Article is based on a paper presented at the ISA/76 International Conference, Houston.

Predicting Control Valve Noise from Pipe Vibrations

C. REED

Field measurement of control valve noise is often impeded by high ambient sound levels in the environment where the valve is located. By measuring pipe wall vibrations, it is possible to calculate the noise generated by a valve without resorting to test chambers or shutting down the rest of the piping system.

Control valves are one of the few sources of over 100 dBA sound levels in chemical, petroleum and power plants. Although many different methods are available to cope with control valve noise after valves and piping are installed (Ref. 3.8), it is sometimes difficult to determine exactly which valve in a complex piping system is responsible for a particular noise level.

The ability of noise and vibration to travel through piping and radiate to the air over large distances from a valve has a significant effect on the sound field around a piping system. Rather than acting as a point source, where sound pressure reduces by 6 dBA per doubling of distance, valve-piping systems can approach the field characteristics of a line source (Ref. 3.9). In a line source, a 3 dBA drop in sound pressure occurs with each doubling of distance. Coping with valve noise requires knowledge of the noise source, but measuring sound levels in installed systems is complicated by the effects of other noise sources in the area. Sometimes these noise sources can be shut down; however, compressors and other integral parts of the system cannot always be turned off for testing purposes.

Consequently, it may be desirable to measure the sound field indirectly by monitoring the wall vibrations of piping associated with a particular valve. Wall vibrations, induced by random acoustic pressure fluctuations generated by the control valve, are re-radiated by the piping exterior into the surrounding environment. This article describes a method to predict control valve noise based on excitation of the pipe wall. Generally, the metod was developed around three basic assumptions:

- Pipes radiate sound in the manner of a line source into an acoustically "free field." This assumption is valid for a typical measurement location 3 ft downstream of the valve exhaust plane and 3 ft perpendicularly from the pipe axis.
- Sound propagates virtually unattenuated over the lengths of pipe evaluated (6 ft). This can be proven using References 3.10 and 3.11, where the sound attenuation ΔIL within a pipe can be represented by:

$$\Delta IL = -2.65 \ \alpha L_p$$

$$\alpha \ (\text{for air}) = 9 \times 10^{-5} \ \frac{\sqrt{f}}{R_p}$$

  (See Nomenclature list in Table 3-4 for description of variables.) Using these equations, the worst case sound attenuation along a 6-ft length of 2-in. pipe at 12,500 Hz is 1 dB.
- The frequency domain over which the method has been verified is limited to the region between the

Table 3-4. Nomenclature

A	area
a	acceleration
c_o	speed of sound in air
c_s	speed of sound in steel
D	diameter
d	distance from source to receiver
f	frequency
g	gravitational constant
I	acoustic intensity
L	length
P	sound pressure
P_{ref}	reference pressure
R_p	pipe radius
t	pipe wall thickness
V	particle velocity
W	acoustic power
α	attenuation function
ω	angular frequency
ρ	density of air

coincident and pipe ring frequencies. The coincident frequency f_c occurs when the phase velocity of sound in the pipe wall is equal to the velocity of sound in the fluid:

$$f_c = \frac{c_o^2 \sqrt{3}}{\pi t\, c_s}$$

The ring frequency is the first circumferential pipe mode; i.e., the frequency at which the pipe circumference is equal to the wavelength of sound. Ring frequency is expressed:

$$f_r = \frac{c_s}{\pi D_p}$$

This simplification was made because dominant sound pressure levels are always found in this frequency domain.

DEVELOPING THE MODEL

From fundamental acoustics (Ref. 3.12), particle velocity V is directly related to sound pressure P via the acoustic

311

impedance, p_0c_0 in the form:

$$P = p_0c_0 V/g \qquad (3.6)$$

Assuming that the particle velocity at the pipe wall is equal to the wall vibration velocity:

$$V = \frac{a}{\omega} = \frac{a}{2\pi f} \qquad (3.7)$$

Substituting this value for V in Equation 3.6:

$$P = \frac{p_0c_0 a}{2\pi f g} \qquad (3.8)$$

where a is acceleration in ft/s^2.

Acoustic intensity I is defined as the sound-power per unit area, or $I = W/A$. Acoustic intensity can also be defined as the pressure squared, divided by the acoustic impedance. Equating the two intensity relationships results in:

$$I = \frac{W}{A} = \frac{P^2 g}{p_0c_0} \qquad (3.9)$$

Sound power can now be expressed as:

$$W = \frac{P^2 A g}{p_0c_0} \qquad (3.10)$$

Consequently, the total sound power radiated by the pipe can be given as:

$$W_p = \frac{P_p^2 A_p g}{p_0c_0} \qquad (3.11)$$

Similarly, the power received at distance d is:

$$W_d = \frac{P_d^2 A_d g}{p_0c_0} \qquad (3.12)$$

Since the sound power radiated by the pipe must equal the sound power received at distance d:

$$\frac{P_p^2 A_p g}{p_o c_o} = \frac{P_d^2 A_d g}{p_o c_o} \qquad (3.13)$$

or

$$P_d^2 = \frac{P_p^2 A_p}{A_d} \qquad (3.14)$$

Substituting $A_p = \pi D_p L_p$, $A_d = 2\pi d L_d$ and assuming that the length of the radiating area is equal to the length of the receiving area ($L_d = L_p$), Equation 3.14 becomes:

$$P_d = \frac{P_p^2 D_p}{2d} \qquad (3.15)$$

From Equation 3.8, $P_p = p_o c_o a / 2\pi f g$. Substituting for P_p in Equation 3.15:

$$P_d^2 = \frac{p_o{}^2 c_o{}^2 a^2 D_p}{8\pi^2 d f^2 g^2} \qquad (3.16)$$

Substituting values for p_o, c_o and g, and assuming that $d = 3$ ft (receiver is 3 ft from the pipe axis), Equation 3.16 becomes:

$$P_d^2 = 7.1 \times 10^3 \, a^2 D_p / f^2 \qquad (3.17)$$

where pressure is expressed in microbars.

The sound pressure level (SPL) is related to the sound pressure P_d by:

$$SPL = 10 \log \frac{P_d^2}{(P_{ref})^2} \qquad (3.18)$$

where $P_{ref} = 2 \times 10^{-4}$ microbars.

Substituting Equation 3.17 for P_d^2 in Equation 3.18 gives:

$$SPL = 10 \log (1.78 \times 10^{11} a^2 D_p / f^2) \qquad (3.19)$$

Acceleration a, however, is in ft/s². For acceleration in g's, Equation 3.19 becomes:

$$SPL = 10 \log (1.82 \times 10^{14} a^2 D_p / f^2) \qquad (3.20)$$

An expression can be developed for any pipe size by substituting for pipe diameter D_p:

2-in. pipe: $SPL = 10 \log (3.1 \times 10^{13} a^2 / f^2)$ $\qquad$ (3.21)

4-in. pipe: $SPL = 10 \log (6 \times 10^{13} a^2 / f^2)$ $\qquad$ (3.22)

8-in. pipe: $SPL = 10 \log (1.22 \times 10^{14} a^2 / f^2)$ $\qquad$ (3.23)

TEST CONFIGURATION

To verify the model, tests were conducted on 2- and 4-in. schedule 40, 80 and 160 pipes and 8-in schedule 20, 40 and 80 pipes. Tests were conducted inside a specially constructed acoustical chamber, Fig. 3-25, so that sound and vibration data could be obtained simultaneously. Since control valve noise is radiated from the control valve itself, the control valve noise source was mounted just outside the chamber. Preliminary tests verified that sound levels measured with the valve inside the chamber were indeed equivalent to those measured with the valve outside the chamber.

Fig. 3-25. Tests conducted inside this acoustical chamber verified that calculated sound levels were the same as those measured acoustically.

Vibration measurements were made at several locations along the pipe length. Special care was taken in accelerometer mounting since high frequency data was required. A B&K 4339 accelerometer rigidly mounted to the pipe wall gave the best results. Sound level was measured simultaneously at the downstream end of the pipe 3 ft from the pipe axis. A Masoneilan Camflex valve, operated in the flow-to-open mode over pressure ratios from 1.7 to 10 was used as a sound source.

Both the sound and vibration data for all test configurations were recorded on a 2-channel Kudelski "Nagra IV" tape recorder for subsequent analysis.

A CLOSE MATCH

Reliable data over the frequency range of interest (from the acoustical coincident frequency f_c to the ring frequency f_r) was easily acquired for the 4- and 8-in. pipe sections. For the 2-in. pipe, where the ring frequency ranges from 26,520 to 28,090 Hz, data could not be obtained over the entire frequency domain. For this pipe size, an upper frequency limit of 12,500 Hz was imposed due to the effect which the greater radius of curvature of the 2-in. pipe had on the accelerometer mounting as well as the practical acoustical data acquisition limit of 16,000 Hz. This was not a serious restraint in the calculation of A-weighted sound levels, however, due to biasing of the A-weighted filter network against frequencies above 10,000 Hz.

Comparing sound pressure levels measured directly to those calculated from pipe vibration resulted in very close agreement, similar in accuracy to that expected from the compound measurement accuracy of acceleration and sound. The standard deviation of the difference between directly and indirectly obtained A-weighted sound levels was found to be 3.05, 1.58 and 1.98 for the 2-, 4- and 8-in. pipes, respectively. The higher standard deviation for the 2-in. pipes reflects the

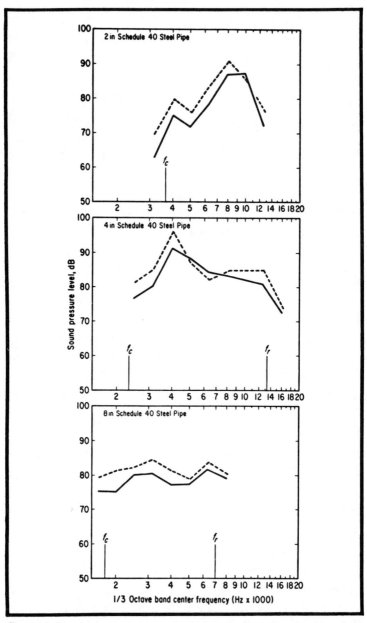

Fig. 3-26. Results for three sizes of schedule 40 steel pipe show the close correlation between actual sound level measurements and calculated values (dotted line). A-weighted sound levels were measured between the coincident frequency f_c and ring frequency f_r for the 4- and 8-in. pipe, but the ring frequency for the 2-in. pipe was above the practical acoustical data acquisition limit of 16,000 Hz.

technical measurement problems discussed previously. Typical comparisons of directly measured and pipe acceleration-based sound pressure levels and sound levels for the three pipe sizes tested are shown in Fig. 3-26.

No significant change in accuracy was found between the different pipe schedules of any size, or with any configuration at various flow rates, Fig. 3-27. This last point is significant since it implies that for the frequency domain of interest, random sound pressure within the pipe dominates over random fluid pressure fluctuations. It is not known whether this would apply for low noise valve designs where the amplitude of the sound pressure inside the pipe would be reduced by 20 dB or more.

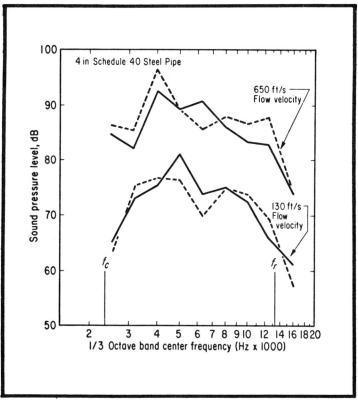

Fig. 3-27. Changes in flow rate have little effect on the accuracy of the model. Dotted lines show calculated sound levels compared to actual sound levels measured in the test chamber at two different flow rates.

APPLICATION NOTES

Successful field use of this method requires proper use of vibration transducers. Because accelerometers are designed for use over specific temperature and frequency regions, as well as orientation with respect to the driving force, it is necessary to use the proper accelerometer for a particular application. The most critical factor is accelerometer mounting. Since high frequency data is required, the transducer should be mounted rigidly to the pipe, with its mounting face flush against the pipe wall.

Installation can be accomplished by welding small metal blocks to the pipe, then drilling and tapping the blocks to accept a mounting stud. The accelerometer can then be screwed directly on the stud, with good face-to-face contact. Simple mounting methods, such as double-back tape or clamps, are inadequate (Ref. 3.13). For other mounting methods, see MIL-STD-740B.

In some applications, pumps and similar equipment can produce vibrations in a different frequency than those generated by control valves. It is also conceivable that other control valves could contribute to pipe wall vibrations. Little is known about the effect a downstream control valve can have on fluid-borne sound generated by an upstream source. Although vibration measurements are often easier to make than sound measurements, engineering judgment is still required.

References

3.8   Arant, J. B., "How to Cope with Control Valve Noise," *Instrumentation Technology*, March 1973, pp. 37-43.

3.9   Scull, W., "Control Valve Noise Rating: Prediction vs Reality," *Instrumentation Technology*, October 1974, pp. 46-49.

3.10   Kinsler, L. E. and Frey, A. R., *Fundamentals of Acoustics*, John Wiley & Sons, New York, 1962, p. 225.

3.11 Ibid, pp. 239-241.

3.12 Ibid, Chapter 1.

3.13 *Handbook of Noise Measurement*, General Radio Co., West Concord, MA, pp. 209-211.

CARLETON REED is a project engineer at Masoneilan International, Inc., Norwood MA.

Section IV—Control Electronics

This final section offers nine chapters about control electronics. These represent a cross section of subjects—from "Solid-state vs. Electromagnetic Relays" to "Fiber Optics for Data Transmission." Specific solutions to problems are provided sometimes as in "Automatic Bridge Balancing Circuit." Others, like "Accuracy in Signal Conversion," discuss theoretical aspects of electronic control circuits.

In "Adapting Electric Actuators to Digital Control," the use of digital-to-analog converters for updating electric valve actuators to digital actuator valve systems is described. The use of incremental encoders, LVDTs, and absolute optical encoders for position monitoring also is covered.

A balanced selection of the latest information pertaining to control electronics is provided.

Solid-State vs Electromechanical Relays

N. REYNER

Numerous factors must be considered when deciding whether to use solid-state relays (SSRs) or conventional electromechanical relays (EMRs). Many users currently have questions as to what characteristics of SSRs should be evaluated and how such parameters can be compared to those of EMRs. The search for the answers to these questions, complicated by the lack of information published on the key parameters on each type of device is by no means straightforward.

When deciding which technology to utilize in a particular application, there are a number of factors which must be considered, some of which are obvious, some not so obvious. Table 4-1 shows the regions where, in terms of load-handling capabilities, either an SSR or an EMR can apply.

Specific design requirements which dictate the selection of electromechanical or solid-state relays in the regions of overlap are:

- cost target
- reliability
- available space

Table 4-1. EMR and SSR Capability Overlap

Input voltage	AC load current			
Output voltage	1-50 mA	50-100 mA	0.1-0.75 A	0.75-25 A
				SSR (molded package)
			SSR (DIP)	
				← 5 A Medium-power clapper
				Solenoid-actuated
	AC optical coupler			
		Miniature general-purpose clapper		
<240 Vac			Large general-purpose clapper	
	← 10 mA			
DC	Miniature reed (<120 V)			

- logic compatibility
- temperature range
- RFI specs
  fault isolation
- environment
- surge current
- operating rate
- switching time

Depending on design specs, the use of either a solid-state relay or an electromechanical device will be dictated. For the regions of interchangeability of SSRs and EMRs, the following specific requirements immediately narrow the selection process to solid-state relays:

- compatibility with CMOS and TTL logic
- high RFI rejection
- operation in an explosive environment
- surge current protection of load
- high-speed operation where switching times are less than 5 ms for power handling and 1 ms for signal handling
- switching operations which exceed 200,000 cycles for power handling and 5 million cycles for signal handling

or to electromechanical relays:

- load voltage under 10 V
- compensation for inadequate heat sinking
- safety, line isolation
- minimal unit machine cost.

For those applications where neither an SSR nor an EMR is mandated by design specifications, factors such as cost, reliability, available space, environment, usage rates, control signal characteristics, to name a few, should be considered collectively. However, it should be emphasized that often the final decision may be ultimately influenced by factors other than device specifications, some of which follow:

- selection of frequently used parts to reduce total component count and achieve larger procurement base
- requirement to use company standard parts
- design philosophy edict to utilize a particular technology
- confidence of design personnel in an established technology.

SSR DESIGN CONFIGURATIONS

SSR circuits can be divided into three distinct, but interdependent sections: the input stage, the control amplifier and the output power stage. Input stages typically consist of reed relays, transformers and optical couplers. Of these, the optically coupled input stage is the most popular because of its speed, recent design improvements and low cost. The reed relay input stage is rapidly declining in usage since its switching speed is too slow to enable zero crossover switching; it also has the mechanical reliability limitations inherent in electromechanical devices. Transformer coupling appears to hold limited promise, except for special applications, because of its cost and its capability of causing signal interference.

Control amplifiers are generally similar, particularly those devices which have ratings in excess of 1A. The signal from the input stage is typically amplified and ANDed with the zero crossover of the line voltage to zero switch the output stage. Proper lead and lag network compensation is required to ensure that the firing window is confined to the region between zero crossover and the turn-on threshold level of the device. Line synchronization is accomplished with a diode bridge connected across the output thyristor to provide rectified AC voltage to the amplifier as the source of power and as a precise definition of the point of zero crossover of the line voltage.

Two types of output stages are common, one of which uses a single triac to switch both halves of the line cycle. The advantages inherent in this approach are simplified circuitry, lower cost and reduced space requirements. The other common approach employs two SCRs in an inverted back-to-back configuration, where only one of the SCRs conducts during any portion of the line cycle. This configuration has such circuit advantages as higher dV/dt, lower gate drive requirements, more positive turnoff characteristics and better heat dissipation. It suffers mainly from cost and space constraints caused by using two thyristors instead of one. One solution is to fabricate the relay from bare SCR chips; however, this approach precludes the use of hermetically sealed devices which are more reliable under power cycling conditions.

RELAY CONFIGURATIONS AND PACKAGING

EMRs come in many package sizes and shapes and in contact arrangements ranging from single form A to multiple form C. SSRs, on the other hand, are currently available only with a single form A switching function.

SSRs are typically fabricated with discrete components which are assembled on a small printed circuit board, where the power stage is mounted on a heat spreader plate. These devices are packaged in a molded or plotted enclosure with their terminations brought out as solderable leads for printed circuit mounting or as connections for chassis mounting. They are primarily confined to the following types of packages, depending on their current rating: oversize DIP—less than 1 A; molded package for PC board mounting—less than 2A; and molded package for heat sink mounting—2 to 40 A. They are available with zero crossover switching, optical, transformer or reed coupling and with SCRs or triacs as output thyristors.

The physical volume occupied by a solid-state relay in relation to its load current rating is indicative of the technology used in its fabrication. Figure 4-1 shows a plot of current switched per package volume as a function of current switched. For comparison, a plot of typical current densities for two-pole electromechanical relays are included.

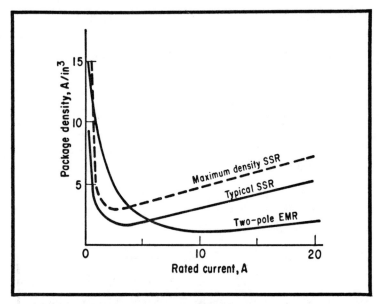

Fig. 4-1. These plots show that, as the rated current handling capacity increases toward 20 A, typical and maximum SSR package density increases accordingly. However the density of a two-pole EMR package remains relatively constant as its rating increases. At lower current ratings, the package density of SSRs increases rapidly as higher levels of integration are utilized.

The actual ranges of package density for available SSR configurations are as follows: DIP devices—8 to 17 A/in.^{3}; molded package for PC board mounting—2 to 4 A/in.^{3}, and chassis mounting—2 to 7 A/in.^{3}. DIP package densities are indicative of the compactness of thick film and pellet construction; however, a practical limit exists due to minimum volume requirements for heat dissipation. All other packages are limited by the physical space required for discrete components and heat dissipation requirements.

COMPARING SPEED, ISOLATION CAPABILITIES

The speed of SSR operation at turnon is primarily limited by triac/SCR response times which are typically less than 5 msec. However, the response time is further increased by the following factors:

- inclusion of a zero crossing network
- amount of time to reach commutation voltage

- addition of a snubber network which must be discharged before load current flows
- use of transformer of reed relay coupling.

SSR turnoff time depends on the portion of the line cycle at which it is initiated, since all thyristors require the instantaneous load current to drop below a holding value before switching off. For all practical purposes, turnoff time should be considered half a line cycle.

Ranges of turnon times for different SSR and EMR configurations are shown in Fig. 4-2, where times for EMRs include bounce time. SSRs are typically at least 10 times faster

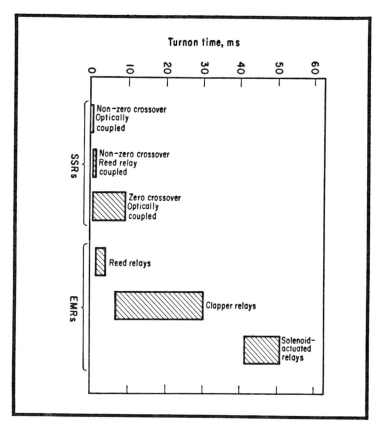

Fig. 4-2. Solid-state relays turn on at least 10 times faster than their electromechanical counterparts. SSR speed at turnon is primarily limited by triac/SCR response times.

Table 4-2. SSR and EMR Isolation Characteristics

Device type	Input/output insulation resistance (MΩ)	Input/output dielectric strength (V)
SSR	1,000-100,000	1,500-2,500
EMR	1,000-10,000	900-2,000

Output off-state resistance (MΩ)
0.006-0.24

than reed relays and 60 to 500 times faster than clapper and solenoid-actuated relays.

The following isolation characteristics are important indicators of a relay's suitability for a particular application: input-to-output insulation resistance, dielectric strength and output off-state resistance. Table 4-2 shows the range of these parameters for typical SSRs and EMRs.

An obvious shortcoming of SSRs is related to off-state resistance, a very important parameter—particularly when safety isolation circuitry is involved. Table 4-3 shows that high current SSRs have leakage currents at rated voltages which are within the region of perceivable pain for humans and where many will experience paralytic effects which may prevent them from letting go of the circuit. This can result in either tissue damage or, if the current path is through the chest, death by asphyxiation.

DRIVE AND EFFICIENCY CONSIDERATIONS

One of the most important features of solid-state relays is their ability to interface directly with all TTL circuits and several types of CMOS buffer driver chips. This feature must be considered when comparing the cost of SSRs and EMRs for applications which require this interface capability since all

Table 4-3. SSR Leakage Levels

Blocking voltage	<1 A	1-10 A	10-40 A
120 V	1-5 mA	4-12 mA	6-20 mA
240 V	1-5 mA	5-15 mA	6-20 mA
480 V	-----	8-10 mA	10 mA

EMRs, with the exception of hybrid relays, need a driver stage. In fact, for high current EMRs, the driver stage necessitates interfacing with power line voltages, since standard coils are designed for minimum drive voltages of 125 V.

Figure 4-3 shows a comparison of typical SSR and EMR input/output characteristics, where pseudo power gain is plotted. The relationship of power gain to the input drive requirements, the blocking voltage and the on-state current for which the relay is rated (on an individual switching pole basis) can be expressed as:

$$Power\ gain\ =\ 10\ \log\ \frac{E_b\ I_c}{E_i\ I_i}$$

where

E_b = maximum blocking voltage
I_c = maximum on-state current
E_t = input voltage
I_t = input current

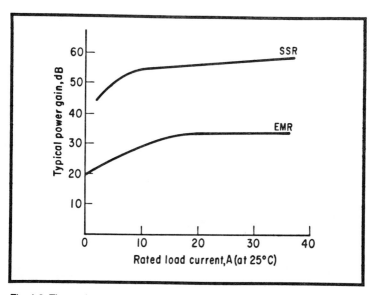

Fig. 4-3. These plots indicate that the input power required by SSRs is typically 300 times less than that required by typical EMRs.

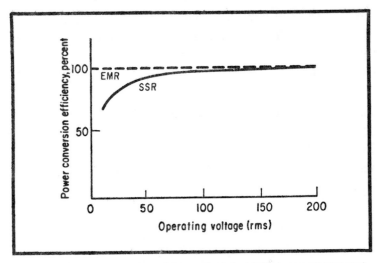

Fig. 4-4. The power conversion efficiency of a typical SSR droops at low operating voltages. This effect is particularly noticeable at low operating voltages because of the voltage drop across the output stage and that of the p-n junctions in the control amplifier which must be exceeded to turn on the SSR.

As illustrated in Fig. 4-3, SSRs require input power levels that are typically 300 times lower than those for comparable EMRs.

Although most users of SSRs are aware of the need to provide adequate heatsinking to prevent overheating of the device, the cause of this problem—power loss in the output stage—has a more subtle effect. Since output thyristors in SSRs are not perfect switches, they develop a voltage drop which is referred to as commutation voltage.

In addition to causing triac heating, the commutation voltage drop reduces the power conversion efficiency of the controlled circuit. As shown in Fig. 4-4, this effect becomes more pronounced particularly at lower operating voltages. For electro-mechanical relays, the conversion efficiency is essentially equal to 1.

We have seen specification sheets for SSRs that include only a maximum voltage limit for the device. However, since voltage drops also develop across several p-n junctions in the SSR, the switching threshold voltage—the lowest voltage at which an SSR can turn on—is often 4 to 5 V. Therefore, due to the effect of the commutation voltage and the switching threshold voltage, data sheets should carry a lower voltage limit.

FAILURE RATES AND COSTS

A theoretical failure-rate analysis of a typical SSR circuit and an equivalent EMR is plotted in Fig. 4-5A for the following conditions:

- normal lab environmental conditions
- cycle rates: less than 10/h
- output stage of SSR: hermetic triac derated by 20 percent
- SSR optically coupled
- EMR: general-purpose clapper configuration.

Figure 4-5A shows that EMR and SSR failure rates for low cycle rates are very nearly the same; EMRs may be slightly lower. SSR failure rate is influenced primarily by the output thyristor, in this case a triac. Over 50 percent of the total failure rate, where $\lambda = 0.1$ to 0.2 percent per 1,000 h, is theoretically associated with this stage. This figure agrees well with experimental test results we have seen, but it does not necessarily agree with supplier specifications.

As the cycle rate increases, the failure rate of EMRs increases rapidly. Figure 4-5B shows the predicted failure rate normalized by the low-frequency failure rate λ_{lf}. The effect of cycling is not as dramatic on SSRs as it is on EMRs if the temperature of the output triac stage case stays within a safe thermal cycle, i.e., less than 25°C.

Estimated large-volume prices were tabulated for typical SSRs and EMRs as a function of the rated output current. A smoothed curve of these values, plotted in Fig. 4-6, provides an additional basis for comparison. The curve is normalized to the cost of an electromechanical relay with two form A contacts, each of which can switch the specified current. Although

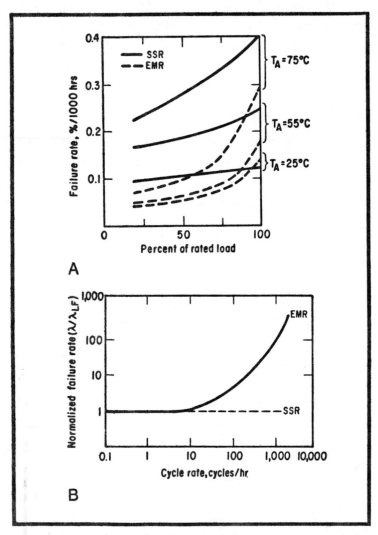

Fig. 4-5. A) The failure rates of EMRs and SSRs remain nearly the same at low-frequency operating cycles. B) On-off cycling affects an SSR much less than it does an EMR if the case temperature excursion of the output stage does not exceed 25 °C. Over 50 percent of the failure rate can be attributed to the SSR output thyristor.

there is some degree of variation in the relative costs of the two devices that have the same rated contact current, a convenient estimator for approximate calculations is that a DPST electromechanical relay costs about 10 percent of an equivalent solid-state switching network, i.e., two SSRs which have a common input signal. This comparison assumes that the heatsinking required for the SSR can be accommodated by mounting it on a chassis of suitable size so that no additional component costs are incurred.

STATUS OF SSR, EMR TECHNOLOGIES

The remainder of this article represents the opinions of several people who have analyzed the SSR and EMR suppliers from the viewpoint of a customer who is concerned about:

- technology utilized
- design safety margins
- internal product design
- assembly techniques
- production process flow
- inspection levels
- technology trends for the future.

The solid-state relay industry is more modern than the EMR industry from a technology viewpoint; however, it has not really capitalized on the fabrication process available to it. For example, only two suppliers of all those surveyed

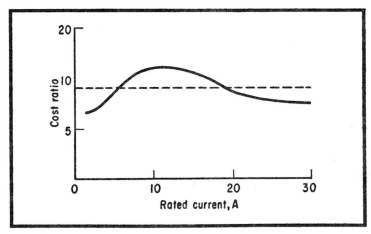

Fig. 4-6. To obtain this relative cost curve, estimated large volume prices were tabulated for typical SSRs and EMRs. The curve, normalized to the cost of an EMR with two form A contacts, shows that the price of a DPST EMR is about 10 percent of an equivalent SSR configuration.

employed semiconductor integration techniques; in both of these cases, the technique was thick-film integration. Some suppliers have demonstrated the practicality of soldering the output thyristor in chip form to copper circuitry and then attaching it to a heat spreader. However, no one was found who either is now applying or has definite plans to apply monolithic integration techniques to an SSR product line. Unless it takes advantage of advanced types of integration, which appears possible only if a large semiconductor house were to enter the field, the SSR industry will not provide broad-based competition with electromechanical devices.

The EMR industry, being generally conservative, has a myriad of standard configurations to meet the needs of most customers, and has an apparent willingness to customize its products to the specifications of volume users. Present relay technology essentially uses proven techniques which fabricate the product in an assemble-and-adjust fashion, where process automation is minimal. Since most electromechanical relay configurations were developed and originally tooled when labor costs were lower, their adaptability to automated production was not a dominant issue.

The major innovations anticipated in EMR products will be application-oriented rather than associated with changes in basic physical principles. With the advent of microprocessors for machine control logic, relays are being designed for direct interfacing; relay packages now contain signal buffering and amplification. Reed relay configurations will increasingly be made more mechanically compatible with printed circuit boards. Better plating processes are also expected to extend electrical life and reliability closer to the design targets of solid-state circuitry. Physical miniaturization, such as in the case of the TO-5 case relay, is possible, although at a cost premium not normally acceptable to the commercial user.

Overall, it is anticipated that the EMR market will expand from $375 million in 1975 to $500 million in 1980. This figure is predicated on the assumption that no significant developments will occur in the cost or available configurations of SSRs that will materially reduce the marketability of EMRs. One instance of where this effect has occurred is the impact of inexpensive and reliable optical couplers on the growth of reed relays.

Although expansion from a current level of $28 million at a historical rate of 17 percent until 1980 appears possible for the reed relay, it is obvious that the sale of $25 million worth of optical couplers in 1975 has had a significant effect on the growth of the reed relay market. One large manufacturer has reduced reed usage by 10 percent on modern versions of his product, while the number of potential applications actually increased.

In summary, EMR technology and its related products are presently confined to relatively stable, time-proven bounds. No innovations are apparent which will significantly alter relay configurations, specifications or construction in the near future. The primary emphasis will be placed on controlling production variables which degrade performance. The larger suppliers will continue to integrate vertically into the end-equipment markets.

NOEL REYNER is Chief Engineer at Transmation, Inc., Rochester, NY.

ABCs of SCR Controllers

S. S. KINTIGH

Industrial power control problems associated with contactors can be solved reliably and economically with packages based on silicon-controlled rectifiers (SCRs). SCRs typically control single or three-phase power to such inductive loads as motors and such resistive loads as electric furnaces. They can also produce DC power from AC sources. These semiconductor devices are compact, silent in operation and unaffected by vibration and shock and have large power-handling capabilities. A well-designed and fabricated SCR has no inherent failure mechanisms; when properly applied and protected, it will provide virtually limitless operating life even in harsh environments in industrial plants.

A complete controller assembly contains SCRs, associated electronics, heat sinks and protection circuits. This package accepts standard command and feedback signals (0 to 5 V, 1 to 5 and 4 to 20 mA) which determine the level of the desired output—either current, voltage or power. A single modern controller containing several SCRs can handle several megawatts of power and provide long and reliable operation.

HOW AN SCR CONTROLLER WORKS

An individual SCR has three terminals: a cathode, an anode and a gate, Fig. 4-7A. It permits current to flow in only one direction and has only two states—on and off. A drive signal applied to the gate lead turns the device on and current flows through the load. Once it turns on, an SCR will revert to the off state only when its cathode-to-anode current decreases below a value called the holding current. SCRs can remain in the off state even though the applied potential may be several thousand volts; in the on state, they can pass several thousand amperes.

SCR operation can be explained in terms of relay and transistor switching circuits, Figs. 4-7B and 4-7C. When the switch closes in the relay circuit, Fig. 4-7B, the coil energizes and the contact closes. The relay keeps the coil energized and current flows through the load. The electronic equivalent, Fig. 4-7C, more closely depicts SCR operation. A momentary gate signal which turns on transistor Q_1 permits current to flow from its emitter to the base of Q_2, to turn on this transistor as well. The collector current of Q_1, holds Q_1 on, even though its gate signal is no longer present. Both transistors remain on until the current diminishes to a very low value.

Figure 4-8 illustrates some of the common SCR controller and load configurations. Referring to the single-phase back-to-back configuration, Fig. 4-8A, it can be seen that the magnitude of the voltage applied to the load can be varied from 0 to 100 percent with nearly infinite resolution by varying the time within an electrical cycle that the SCRs are on. When the turn-on time is varied within an electrical cycle, the SCRs are called phase-fired. Phase firing provides extremely fine resolution and can control fast processes.

Another mode of control, termed distributed zero crossover, involves varying the duty cycle ratio of full cycle power applied to the load. For example, if the SCRs are turned on every other electrical cycle, the power applied to the load

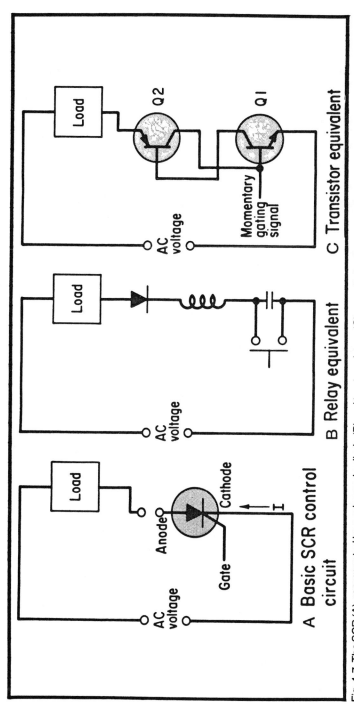

A Basic SCR control circuit

Anode

Cathode

Gate

I

B Relay equivalent

Load

AC voltage

C Transistor equivalent

Load

Q2

Q1

Momentary gating signal

AC voltage

Fig. 4-7. The SCR (A), represented by a relay and a diode (B) and two transistors (C), permits load current to flow after a gating signal turns it on. It remains on until the current decreases below the holding current level. Resistive loads can be direct-coupled or, for impedance matching, transformer-coupled to the SCR anode.

would be 50 percent of the power available if power were applied continuously. The advantage of distributed zero cross-over operation or zero voltage switching is that it essentially eliminates radio frequency interference often encountered when the applied power or voltage is controlled by phase firing.

Zero crossing control also eliminates the reduction of power factor which occurs in phase control. The power factor that exists in phase-angle control can be computed by dividing the load voltage by the actual line voltage. For example, if the SCRs are conducting and the load drop is 240 V with the line voltage at 480 V, the power factor is 0.5. Typically, the zero crossover mode is associated with heating element control for ovens and furnaces where the thermal load inertia is relatively high. This mode of operation can be used successfully in any application that would normally include a contactor.

COMMON CONTOLLER CONFIGURATIONS

The three-phase, two-leg controller configuration, Fig. 4-8B, is the least expensive version because it contains only four SCRs. It operates best in the distributed zero crossover mode and is commonly employed for heater control. The two-leg system can be phase-fired, although under some conditions the SCR cathode potential may be reversed when the gate signal is applied. This condition causes excessive SCR leakage and temperature.

The three-phase hybrid system, comprised of three SCRs and three diodes, as shown in Fig. 4-8C, is normally the least expensive phase control assembly. However, it is more susceptible to the introduction of DC components than the other phase control configurations; therefore it is not recommended for use with inductive loads. The three-phase, six-SCR in-line approach, Fig. 4-8D, can be operated in either the phase-fired or the distributed zero crossover mode. This arrangement is probably the most common three-phase SCR

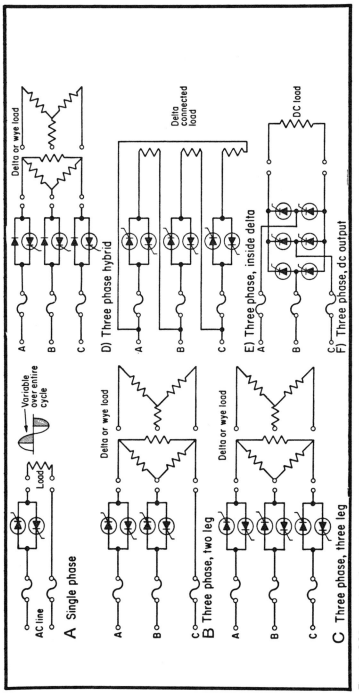

Fig. 4-8. SCR controllers for single- and three-phase industrial applications are usually configured as shown in (A) through (E). The AC-to-DC conversion circuit (F) is commonly employed in metal processing applications.

343

controller approach. It can be used for both inductive and resistive loads in delta or wye configurations.

The three-phase, inside delta scheme, Fig. 4-8E, can supply appreciably more power for a given size SCR assembly than can any of the other systems. Power to a three-phase load is of course equal to the phase-to-phase voltage times the line current times the square root of 3. In the three-phase inside delta arrangement, the current that the SCRs are required to conduct is equal to the total power dissipated in the load divided by three times the line voltage. Therefore, the inside delta, for a given SCR capability, can provide approximately 1.73 times the power than any other of the arrangements discussed.

Metal processing such as aluminum refining and tungsten fabrication often requires DC current or voltage. DC current and voltage can be obtained from AC sources by several SCR conversion schemes; Fig. 4-8F shows a common one. In this application, the SCRs regulate the power to the load and provide the rectification necessary to obtain low ripple DC with 0 to 100 percent control.

Another SCR application is in wire annealing, where load shorts occur frequently. This application employs an electronic fuse that substitutes for a conventional fuse by interrupting the gate signal to the SCR if the load current exceeds a preset value. Obviously this circuit must be very fast and reliable.

New SCR capabilities are continually being developed and the number of applications continues to grow. However, not all SCR applications have been successful. In the author's opinion, this is not the fault of the controller itself, but is due instead to the manner in which it has been applied and protected. It is the vendor's responsibility to make these determinations, because the user cannot be expected to be aware of all equipment parameters required for successful operation.

HARDWARE SELECTION GUIDELINES

Since approximately 40 different parameters may constitute an SCR controller's specifications, the following factors are given as a guide to the primary considerations in selecting a particular unit.

Voltage rating—SCRs and associated components such as transformers must be capable of withstanding at least 40 percent more than the peak value of the line voltage, since line voltage transients of this magnitude can occur in industrial applications. Failure to allow for high line voltage, even if it is only transient in nature, can cause SCR failures.

False firing protection—SCR specifications include limits on the maximum rate of change of voltage that can be withstood. Faster changes can trigger the SCR inadvertently which, in the case of inductive loads, may result in the immediate failure of fuses or the SCRs themselves. In addition, fast voltage changes can damage the SCRs just enough to produce mysterious failures at a later time. Such transients can make one controller track another SCR controller. This condition may occur because SCRs turn on extremely fast and can place voltage transients on the power line; hence it is possible for one SCR controller to turn on a second SCR controller. A dv/dt protection circuit, consisting of an RC filter, is essential in all SCR controllers.

Current capability—The current handling capability of an SCR is dependent upon the type of heat sinking (air or liquid) provided. Operating at excessive temperatures will cause SCR failures. An SCR controller intended for industrial application should have a thermostat on its heat sink to signal when the temperature exceeds a maximum allowable value. The vendor should be able to state the safety factor of the actual current capability vs the specified current rating.

Fusing—SCRs will fail due to load shorts just as mechanical contactors will. SCRs have ratings specified in terms of I^2T and I surge to permit the proper selection of fuses. For adequate protection, fuses must have lower ratings than the associated SCRs.

Gate drives—The drive circuit that provides the gating signal to the SCR must meet very demanding requirements since SCRs can be worn out if they are not properly turned on. Improper turn-on creates small hot spots within their semiconductor material. Successive failures of this type will ultimately cause the entire SCR controller to fail. Improper turn-on can result from voltage transients or from insufficient gate drive.

The ideal gate signal is one that provides a current independent of the SCR gate impedance and that has a rise time less than 1 μs. It is essential that an SCR be turned on by a current source, which provides the proper gate current regardless of gate impedance. It should also provide a small current called a "back porch" which trails the gating pulse to insure that the SCR is turned on properly when used with inductive or transformer loads. In addition, the gate drive circuitry must be designed to insure that transients on the power line cannot be coupled or passed through it to cause a false turn-on of the SCRs.

Timing—For inductive or transformer-coupled resistive loads, it is extremely important that DC voltage or current is not introduced. The low DC impedance of inductive or transformer loads permit excessive current to flow and this can harm SCRs. This situation arises frequently with phase-fired controllers when a transformer is re-energized after a power interruption. It is quite possible that the flux in the transformer core may be previously set so that when power is restored, the established phase angle produces almost immediate saturation and excessive current flow.

A properly designed controller monitors a loss of one phase of incoming power or power interruptions, and automatically starts and sequences the gating of the SCRs to avoid transformer saturation. In addition, the controller automatically adjusts for component changes in the gating circuits so that the DC voltage across the load is kept to an absolute minimum.

Feedback/command signals —It is desirable to be able to manually adjust the controller output with a potentiometer, and to have the controller accept any of several standard input signals (0–5 V, 1–5 mA, 4–20 mA, 10–50 mA, etc). With an appropriate plug-in feedback transducer, the output of the controller could be changed in form: voltage, current, power or speed. The controller should be capable of adapting to these various requirements. Preferably, the vendor should have these transducers as a part of his product line so that the "marriage" of the controller to the feedback transducer is the vendor's responsibility.

TIPS ON SERVICING, COST AND VENDOR KNOW-HOW

All devices, whether they are mechanical or electrical, can fail and, in the author's opinion, they will do so when needed the most. A well-designed controller must have circuit boards, fuses and SCRs that can be quickly changed, preferably by the user rather than the vendor's field service personnel.

The price of SCR controllers will vary widely depending upon the application for which they are originally designed. A purchaser should be wary of the vendor whose price is either excessively high or low.

The application of an SCR controller requires experience. There are numerous pitfalls waiting for the inexperienced vendor. An established vendor very likely has designed his controller to prevent its misapplication, and has gained working knowledge essential to successful usage. An experienced SCR controller vendor will provide applications engineers and field service personnel to assist the customer.

STANLEY S. KINTIGH is Engineering Manager of the Process Systems, Div. of Research, Inc., Minneapolis. Article is based on a paper presented at the ISA/75 Conference, Milwaukee.

Fiber Optics for Data Transmission

D.N. WILLIAMS

Fiber optics technology is now mature enough to be implemented as interface cabling for various systems including aircraft, telephone exchanges, data communication networks and many others. Fiber optic interconnections offer highly significant advantages over conventional wire cabling: noise immunity, total electrical isolation, and reduced size and weight.

Transmitting data by light through optical glass or plastic fibers is potentially far superior to communicating information over conventional electrical wires. Such fibers offer several advantages, including wide signal bandwidth, electrical isolation, no crosstalk, interference immunity and lightweight, low-volume cabling.

Fiber optic components come in several forms: 1) single fibers or light pipes which conduct light signals from a small modulated source to a photodetector, 2) light conduits, comprised of several fibers randomly placed on a bundle, which perform the same function as single fibers, 3) image conduits formed from groups of fused fibers, and 4) imaging faceplates and lenses.

Table 4-4. Optical Fiber vs. Conventional Cables

Parameter	Fiber optic	Coaxial	Twisted pair
Low-level crosstalk	•	•	
No measurable crosstalk	•		
RFI/EMI/EMP immunity	•		
Total electrical isolation	•		
No sparks/fire hazards	•		
No short circuits/loading	•		
No contact discontinuity	•		
Withstand temperatures to 300 °C	•	•	•
Withstand temperatures to 1,000 °C	•		
Signal bandwidth to 1 MHz (300 m)			•
Signal bandwidth to 20 MHz (300 m)		•	
Signal bandwidth >200 MHz (300 m)	•		
Lightweight materials	•		•
Low cost	•	•	•
Vibration tolerant	•	•	•

Table 4-4 compares the characteristics of fiber optic cables against those of their coaxial and twisted-pair counterparts. If no measurable crosstalk is desired between channels in a cable, the only practical choice is fiber optics.

Optical fibers, being made of dielectric materials, provide optimum immunity to radio frequency and electromagnetic interference (RFI/EMI). Since they neither pick up nor radiate signal information, such fibers offer greatly improved electromagnetic compatibility (EMC) over wire cable systems. Because of their dielectric properties, they are also immune to electromagnetic pulses (EMP), a form of interference often associated with high-power explosive detonations that can produce damaging currents in conventional wiring.

A fiber optic transmission system provides total electrical isolation between the sending and receiving terminals, thus eliminating common ground connnections and such associated problems as voltage offsets and ground currents and noise.

Fiber optic linkages present no fire hazards when their fibers are damaged. In addition, no local secondary damage can occur because fiber cables neither produce sparks nor dissipate heat.

Short circuits or circuit loading do not reflect back to the terminal equipment when a fiber optic cable is damaged. In contrast, damage to a wire cable may, in turn, harm the terminal circuits by shorting or grounding them, or by inducing dangerous voltages and currents in the wires which connect to them.

Conventional wiring is greatly affected by connector discontinuities because it needs solid physical and electrical contact at the connector interfaces for optimal signal transfer. An optical interface between the light source or detector and the fiber optic bundle or between two fiber optic bundles provides a signal junction which requires no physical contact. A light signal can pass through a small air gap between the end of each optical fiber and these devices.

Tests have shown that most liquid contaminants found in aircraft which may saturate this interface actually increase the signal coupling. However, bubbles, granular materials and opaque substances can decrease coupling and damage connector surfaces as well.

Typical glass fibers currently available remain stable well beyond 300 °C; special galss cables can withstand temperatures up to 1,000 °C. The maximum operating temperature of fiber cables usually depends on their jacket material or on the epoxy used to secure them in connector ferrules.

The signal attenuation of typical fibers ranges from 2 dB/km to 1,000 dB/km. For example, the signal bandwidth of a 300-m fiber which has 50 dB/km of attenuation and the proper light transmission characteristics is 200 MHz. This limit is primarily a function of the intensity modulation rate limit of the light source. The bandwidth of coaxial cable, independent of signal processing electronics, is limited to 20 MHz for the equivalent diameter and length of fiber optic cable, while a twisted-pair wire has a 1 MHz bandwidth. Fiber

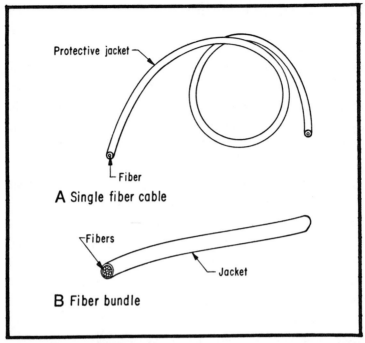

Protective jacket

Fiber

A Single fiber cable

Fibers

Jacket

B Fiber bundle

Fig. 4-9. A single optical fiber (A) or a bundle of fibers (B) is essentially an optical waveguide. The jacket protects the glass fiber or bundle against damage.

optic cables with bandwidths greater than 100 MHz are generally used in laboratory work at present. The potential signal bandwidth of small diameter fibers extends beyond 1 GHz.

SENDING LIGHT DOWN THE PIPE

A single fiber or a bundle of fibers, Fig. 4-9, serves essentially as an optical waveguide. Light propagates through each fiber, as shown in Fig. 4-10, because of the difference in index of refraction between the core and the cladding (glass sheathing fused to the core) which has the lower index. Light propagates at a constant angle with its axis. Fibers that have small diameter cores or low order waveguides permit single mode propagation, whereas multimode propagation occurs in fibers that have larger cores.

A typical fiber optic system, Fig. 4-11, contains a light-emitting diode (LED) as the light source whose intensity varies in accordance with the modulating signal. The intensity of light generated by the LED is proportional to the current flow through it. Depending on the current level required by the LED, its driver may be a single transistor; for high-speeds operation, the driver may be a wide-band amplifier.

All analog AC, DC or synchro signals can be handled either directly, using linear circuits, or with A/D or synchro-to-digital converters. Fiber optic lines are ideally suited to handling such signals as digital data and rf pulses, and they can carry analog signals as well. Although optical fibers are not intended to carry power, they can handle signals which remotely activate local power supplies. Switching functions are handled with ease.

Figure 4-11 shows a pulse signal being transmitted whose logic 1 level turns the LED on, and whose logic 0 level turns it off. The light generated by the LED enters the fiber optic bundle and propagates through it to the photodiode at the receiving end. The level of signal current through the photodiode is proportional to the amount of light impinging on it. This signal is amplified and matched to the following signal handling circuits. The signal-to-noise ratio is determined by

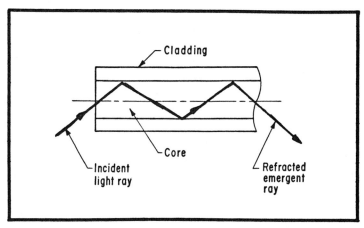

Fig. 4-10. Because the core and the cladding of a fiber, both of which consist of transparent optical materials, have different indices of refraction, a light ray incident upon their interface is internally reflected. Light reflects down the entire length of the fiber at a constant angle with its axis.

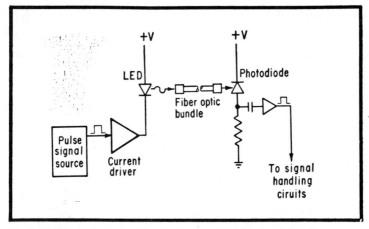

Fig. 4-11. This system transmits pulses by amplifying the input signal in a current driver to turn the LED on and off. Since the light carried by the fiber bundle when the LED is on causes the photodiode to conduct, a pulse develops across the input resistor of the receiver amplifier, corresponding to each input pulse.

the level of the light signal and the noise equivalent power level of the first amplifier.

As shown in Fig. 4-12, the electronics and LED for a bulkhead-mounted digital transmitter can be contained in one module, as can the photodiode and output circuits for the receiver. Both the receiver and transmitter contain transistor-transistor logic (TTL) circuits which can handle a maximum data rate of 10 megabits, Manchester coded and self-clocking, where the minimum pulse width is 50 ns.

A modular fiber optic interface has distinct advantages over interfaces where the semiconductor elements are separated from their respective transmitting and receiving circuits. Temperature compensation for the LED power supply output can be easily achieved when both the LED and output driver circuits are placed in the same package. The importance of proper circuit and package design of the receiver cannot be overstressed, since the photodiode/first-stage amplifier combination establishes the noise, external interference immunity and bandwidth of the interface. Receiver components must be mounted in close proximity; proper shielding is also necessary.

354

Our center is developing a multichannel connector to interface fiber bundles to digital transmitter and receiver modules mounted on printed circuit boards. Analog modules which mount on printed circuit boards are now being developed to handle signals for closed-circuit television and computer-generated CRT displays.

The availability of standard test methods and refined test equipment for such fiber optic components is currently limited. Test procedures and equipment setups to characterize and measure the various parameters of fiber optic components

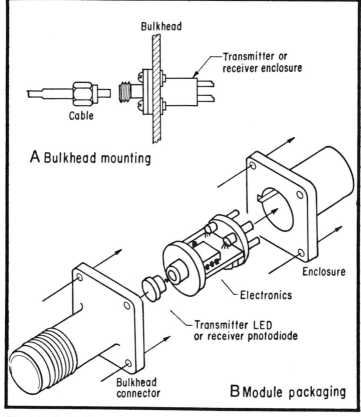

Fig. 4-12. Packaging the transmitter LED with its power supply or the receiver photodiode with its amplifier in one module (A) for bulkhead mounting (B) offers two advantages. The power supply output can be easily temperature-compensated, and the noise input to the amplifier is minimized because of close signal coupling.

and to provide necessary data for circuit and systems designers are also being developed at our center.

INCREASING TRANSMISSION CAPACITY

Since ships, planes, subs and computers include relatively short cables, bundles of fibers can be used for each channel to improve signal coupling at the light source and to provide the high redundancy associated with several parallel fibers. Severing half of the fibers in a bundle introduces only a 3-dB signal loss in that channel, whereas severing a single fiber destroys that channel. For longer land and undersea applications, it is necessary to use each fiber as a separate channel and cables which have highly protective jacketing.

Three types of signal multiplexing may be considered for fiber optic systems. Because of the high bandwidth and EMC advantages of fiber optics, electronic parallel-to-serial or time division multiplexing may be used to transmit many signals in a serial format.

On the other hand, each fiber optic bundle may be subdivided to make each group of fibers a separate channel for parallel data transmission. A single fiber may be used as one channel if the redundancy provided by a group of fibers is not required.

Several light sources which have different wavelength characteristics or colors may be employed to add to the data capacity of a fiber optic cable. Such wavelength or carrier frequency multiplexing allows several channels to be carried on a fiber bundle, subgroups of fibers or a single fiber. At present, five colors in the range from visible red to near-infrared can be used simultaneously for acceptable channel separation. Various combinations of all three multiplexing techniques may be used to greatly enhance the data capacity of fiber optic interfaces.

Crosstalk specs and security considerations for digital communications systems dictate the minimum rise-time value of the transmitted pulses. This constraint often makes it necessary to use cables which contain numerous parallel wires to handle information transfers between two terminals. All of the cabling between such terminals can generally be replaced by one or two fiber optic cables through which parallel data are time-division multiplexed.

Increased cable weight and space savings can be realized by multiplexing greater numbers of signals, thus taking advantage of the inherently wide bandwidth of fiber optic cables. The multiplexing electronics for such cabling, which replaces many line drivers/receivers in a parallel system, consumes less power and contains fewer components.

HIGH FLYING FIBERS

An Airborne Light Optical Fiber Technology (ALOFT) demonstration is being conducted by our center to confirm that fiber optics is a practical, mature technology for use in internal aircraft data signal transmission. The demonstration system includes a tactical computer which has been modified so that its external signals are transmitted to the peripheral avionics over fiber optic cables instead of the original copper wiring. The 115 individual signals transmitted in the original wire navigation and weapons delivery system (NWDS) are time-division multiplexed into 13 serial data channels between the ALOFT computer and the peripheral avionics adapters. A major portion of the weight reduction realized in this system (31.9 lb vs 2.7 lb) is due to multiplexing.

However, since multiplexing is achieved by digital coding at a maximum data rate of 10 megabits, twisted-air wire cannot be effectively utilized as the interface medium in the ALOFT system. Coaxial cable could be used for electronic multiplexing to reduce weight by 8 times, but it would increase susceptibility to EMI. The fiber optics configuration used for signal transmission reduces cable weight by 21 times without increasing susceptibility to EMI normally experienced at these higher data rates. In military aircraft, the high immunity to EMP offered by a fiber optic interface is particularly significant.

Figure 4-13 shows the various ALOFT avionics units, including the tactical compute at right and the adapter boxes. These units contain the signal multiplexing and demultiplexing electro-optic circuits and adapter power supplies which interface to the NWDS avionic units. A 13-channel, bundle-to-bundle connector having individual transmitter and receiver circuits mounted internally on circuit boards comprises the computer interface.

The avionics adapters represent an intermediate step in the ongoing process of incorporating fiber optic interfaces into avionic systems. The next step will be to replace the parallel wire, multichannel line drivers and receivers within each NWDS avionics unit with the multiplexer and demultiplexer circuits in the five adapters. In most cases, the substitution would reduce the component count, power dissipation and space occupied by the interface circuits in each of the NWDS avionic boxes.

A cost analysis is being conducted in parallel to the testing and evaluation of the ALOFT system to establish the potential of fiber optic interfaces in future avionics system. Initial forecasts in this analysis predict that the present costs of components produced by industry should fall to 70 or 80 percent of previous costs each time the initial production volume is doubled. At present, the total cost of cable and connectors is $1,630 for wire avionics interconnections, and $1,030 for fiber optic interfaces. Future costs of the required components should be completely determined by the demand placed on industry to deliver the needed quantities of components.

The conclusions of the ALOFT flight test demonstration and the analysis should answer, in quantitative terms, the questions being posed by avionic systems designers in the Department of Defense and industry about the capability and potential of fiber optics communications. If the results are

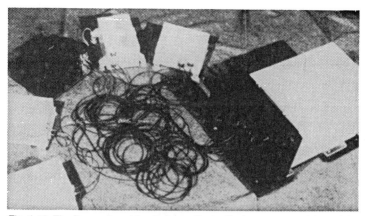

Fig. 4-13. The fiber optic cabling between the tactical computer at right and the adapters for five avionic systems represents 13 channels through which 115 data signals are multiplexed. Twisted-pair wire could not handle the high-speed data resulting from such multiplexing, and using coaxial cables would increase the susceptibility to EMI.

favorable, the ALOFT demonstration should pave the way for a multitude of such advanced avionics systems as point-to-point communication interfaces, avionics data bus systems and fly-by-fiber optical control systems.

FIBER PHONES AWEIGH

Fiber optic cables were chosen to link six stations of a telephone network onboard the 6th Fleet flagship. Figure 4-14 shows one of the telephones in the system which has been deployed for three years. The phone, similar in appearance to a standard Navy sound-powered phone, was designed to operate as part of standard intercom system. The phone system permits three simultaneous, although separate, two-way conversations. The fiber optic cables are shown attached to the top of the phone box and strapped against the bulkhead. In this application, conventional fiber optic cables are employed which have a signal attenuation of 500 dB/km, with a jacket of PVC plastic (no armoring).

The supplemental radio area in the flagship contains the central switching station through which all calls are routed. There are six buttons on the switching station that may be

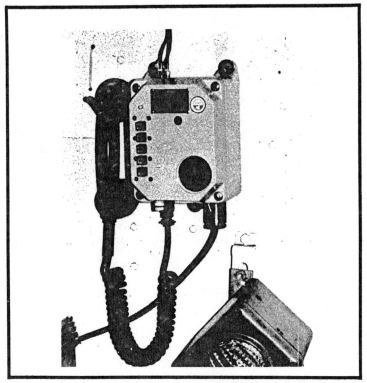

Fig. 4-14. This phone operates as part of a fiber optic intercom system which permits three separate, two-way conversations to take place at the same time. The fiber optic cables are attached to the top of the phone and strapped against the bulkhead.

used to remove any phone from the link. If anyone attempts to use this phone, they get a busy signal, as would anyone attempting to reach the caller.

Optical signals which arrive in the fiber optic bundle from the calling station are detected and converted to electrical form at the central switching station. These TTL logic compat ible signals are processed and directed to the phone station being called. They leave the central switching station again as optical signals in the fiber optic bundles.

At the called station, the optical signals are detected and demodulated to produce an audio signal for the handset re- ceiver. Full duplex operation is obtained by connecting two such signal paths in both directions for each phone link. Our technology program is developing full-duplex communication over a single bundle cable for telephone systems.

REDUCING CABLE WEIGHT AND VOLUME

Studies of the 2,000-ton surface effect ship currently being developed have shown that fiber optic interconnections could be used to reduce the weight of most combat systems, especially of those affected by EMI. The weight of each system's cables represents 10 to 13 percent of its total poundage; for example, cabling for rf communications weighs 1,331 lb or 10 percent of the total weight of that subsystem. Point-to-point connections for 700 signals generated by several monitoring, control, signaling and alarm data systems would weigh approximately 12,500 lb. The weight of such connections can be reduced to 3,000 lb by multiplexing the signals they carry. However by using fiber optic cabling, the weight would be only 750 lb.

A typical fiber optic cable used for system interconnection is ⅛ in. in diameter and weighs 0.005 lb/ft. In contrast, shielded, twisted-pair wire, No. 18, used for the same function is ¼ in. in diameter and weighs four times more than its fiber counterpart.

The compactness of fiber optic cables can prove beneficial when several cables must be terminated in a confined space. For example, approximately 1,700 functions have to enter or leave a shipboard engineering and damage control center. This network would require a termination device to service the cables which occupies 36 cu ft of space and weighs 800 lb. This space and weight penalty can be alleviated by using fiber optic cabling.

Another savings associated with the fiber optics approach results from reduced installation costs. Shipyard estimators allow 0.4 manhours/ft for 24-conductor cables and 6 manhours for all terminations on each end of the cable. If the cables are less than ½ in. in diameter, as are fiber optic cables, the estimators allow 0.125 manhour/ft and 1 hour for each end, with three terminals per connector. The savings in labor costs are estimated to be 68 percent of the labor costs for the present approach to wiring.

DONALD N. WILLIAMS is Project Manager at the Naval Electronics Laboratory Center, San Diego. Article is based on a paper presented at the ISA Aerospace Industries and Test Measurement Divisions Symposium, San Diego, 1976.

Parallel-to-Serial Conversion for PCM Data

S. Z. TAQVI and K. TU

The weight, space and power limitations inherent in complex space probes necessitate minimized data handling logic. Space probe payloads contain pulse-code-modulation (PCM) telemetry systems that process data being gathered by several types of sensors. State-of-the-art particle detectors measuring radiation levels usually require counting or accumulator circuits in the telemetry system to encode their data. Serial accumulators in typical PCM encoders store radiation data generated by the sensors and, at the appropriate readout time, parallel transfer their contents to an output shift register for transmission. To accomplish this transfer of data, logic gates are required for each bit of the accumulator.

With complex payloads, it becomes essential to reduce the gating circuitry for the inherently large number of radiation data accumulators to meet stringent weight, space and power requirements. The method described here employs serial-type accumulators to significantly reduce the amount of logic circuitry needed for parallel-to-serial encoding operations. A space probe telemetry system based on this scheme contains 28 8-bit serial accumulators whose contents are multiplexed into a single PCM stream. The approach works best in PCM systems handling data from low rate (less than 1,000 bps) sensing devices.

CLOCK DOUBLE-DRIVES COUNTERS

A generalized version of this system, Fig. 4-15, has M serial accumulators of N bits each, and an N-bit central accumulator which performs the parallel-to-serial conversions. Asynchronous data pulses from each detector are counted by the associated accumulator, until a readout pulse occurs.

At the leading edge of the readout word pulse (W_L) for the Lth accumulator, the detector input to accumulator L is inhibited. At the same time, a reset pulse sets the central accumulator Q outputs to the logic 1 state and a high frequency clock signal feeds accumulator L and the central accumulator simultaneously. Clock frequency is adjusted so that accumulator L will overflow. When an overflow signal is detected, the high frequency clock is inhibited. The 1's complement of the central accumulator contents is then parallel loaded into the output shift register at the shift register load pulse. Thus, the binary number loaded into the shift register is identical to or a copy of the number stored in accumulator L at the time the detector input was inhibited.

Once the shift register receives the data, accumulator L resumes accepting pulses from the detector and the input to accumulator (L + 1) is inhibited. The readout of the contents of accumulator (L + 1) is initiated and processed the same way as that of accumulator L. The main advantage of this design concept is that parallel-to-serial conversion is performed serially instead of in the conventional parallel way. Since all accumulators share a common gating circuit for data transfers to the central accumulator, the logic gates normally associated with parallel accumulators are unnecessary. For an N-bit accumulator PCM system, the cost reduction factor for the logic hardware is about N.

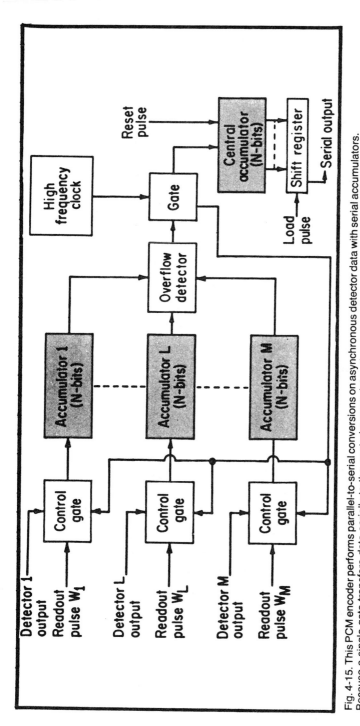

Fig. 4-15. This PCM encoder performs parallel-to-serial conversions on asynchronous detector data with serial accumulators. Because a single gate transfers data serially to the central accumulator, the system does not require the redundant logic associated with parallel data transfer. However, a high frequency clock signal is required to obtain reasonable conversion rates.

365

The frequency of the clock is critical since the time required to overflow any of the N-bit accumulators must be less than the transmission bit period of the PCM system. Clock rate C should be:

$$C \geq 1.2 \; \frac{2^N}{N} \; R$$

where R is the transmission rate and N the number of accumulator bits. For example, if $R = 50$ kilobits/s and $N = 8$, then $C \geq 1.92$ MHz. However, when $R = 1$ megabit/s and $N = 8$, then $C \geq 38.4$ MHz. Thus, the trade-off between the number of logic circuits which can be saved and the logic circuit speed required at a given transmission rate must be considered before implementing this scheme.

A 4-BIT CONVERSION SEQUENCE

Parallel-to-serial conversions performed by this generalized system can be described in terms of a simple 4-bit accumulator system, Fig. 4-16. The Table shows a typical clock sequence and the contents of both a detector accumulator and the central accumulator being incremented by clock pulses. For the following discussion, assume that at the time the asynchronous data input to the accumulator has been inhibited, the accumulator has reached a binary count of 0010 and, at the same time, the central accumulator $\overline{Q}$ outputs are set to the logic 1 state. When the 13th clock pulse occurs, the data accumulator contains all ones; the next clock pulse increments the contents of the data accumulator and causes an overflow to occur. At this time, the most significant bit of the data accumulator changes from a logic 1 to a logic 0; the output of the overflow detector changes from a logic 1 to a logic 0 and inhibits the clock. At this point, the $\overline{Q}$ outputs of the central accumulator are 1101. The Q or complemented outputs, being

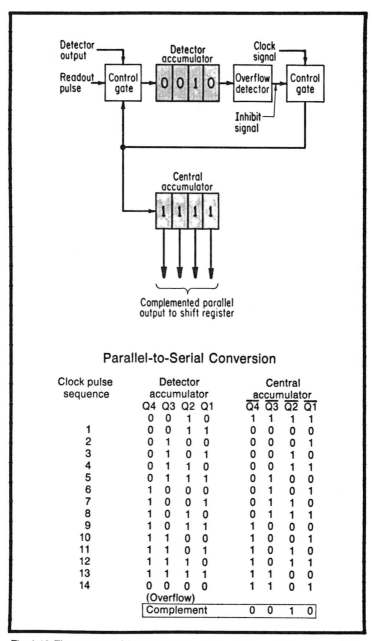

Parallel-to-Serial Conversion

Clock pulse sequence	Detector accumulator				Central accumulator			
	Q4	Q3	Q2	Q1	Q̄4	Q̄3	Q̄2	Q̄1
	0	0	1	0	1	1	1	1
1	0	0	1	1	0	0	0	0
2	0	1	0	0	0	0	0	1
3	0	1	0	1	0	0	1	0
4	0	1	1	0	0	0	1	1
5	0	1	1	1	0	1	0	0
6	1	0	0	0	0	1	0	1
7	1	0	0	1	0	1	1	0
8	1	0	1	0	0	1	1	1
9	1	0	1	1	1	0	0	0
10	1	1	0	0	1	0	0	1
11	1	1	0	1	1	0	1	0
12	1	1	1	0	1	0	1	1
13	1	1	1	1	1	1	0	0
14	0	0	0	0	1	1	0	1
	(Overflow)							
	Complement				0	0	1	0

Fig. 4-16. The contents of the central accumulator and the detector accumulator shown here at the start of a readout period are incremented simultaneously by clock pulses after the detector input has been inhibited. When the detector accumulator overflows, the complement of the number it originally contained remains in the central accumulator.

367

0010, are loaded into the output shift register. This binary number is identical to the number in the data accumulator prior to the inhibit of the detector input.

DR. S. ZAFAR TAQVI is a Staff Manager and KWEI TU is a Staff Engineer at the Aerospace System Div. of Lockheed Electronics Co., Houston. Article is based on a paper presented at the ISA/75 Conference, Milwaukee.

Intrinsically
Safe Data Acquisition
L. PAYNE

Remote data acquisition is an inherent part of any process, whether the application involves closed-loop control, supervisory control, or merely data logging and monitoring. A major problem in many applications involves the amount of wiring necessary to connect remote sensors to a data acquisition system. In some installations, the main cable routes may contain hundreds, perhaps thousands, of cable pairs.

Remote multiplexing is one means of minimizing wiring in new systems and expanding the capacity of older installations. In hazardous locations, remote multiplexing eliminates the need for installing a safety barrier on each sensor lead. However, the remote multiplexer itself often has to be made intrinsically safe in a hazardous location, in most cases through the use of explosionproof or purged enclosures.

Complementary-metal-oxide semiconductor (CMOS) technology has made it possible to build multiplexers and instrumentation components that are intrinsically safe, thus eliminating the need for enclosures. CMOS devices are well suited to applications which require low power consumption, noise immunity, high speed, zero voltage offset and low leak-

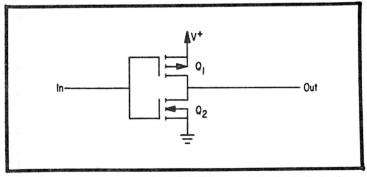

Fig. 4-17. This simple two-transistor CMOS inverter is the basic device from which many complex logic circuits are formed. No current can flow through this circuit except during a change of state.

age current. This article describes basic CMOS circuitry used in the design of intrinsically safe remote data acquisition equipment.

GATES AND INVERTERS

The increasing market demand for intrinsically safe instrumentation is occurring at a time when new semiconductor technology, such as CMOS, is providing devices ideally suited to this purpose. The CMOS logic family has been developed from a versatile technology which allows compact logic gate geometries; these geometries make it possible to place complex logic functions on a single semiconductor chip.

CMOS logic functions are constructed from gates which are in turn built from circuits similar to the two-transistor inverter shown in Fig. 4-17. The two transistors of this circuit are designed so that only one will be "on" (capable of conducting current) for any valid input logic level. The input consists of two FET insulated gates which draw no DC current. When thousands of these devices are interconnected in a CMOS chip, no current can flow anywhere in the circuit except during a change of state. The result is that very complex logic operations can be performed with a supply current of less than 10 → 20 mA at 5 to 15 V. This is an ideal current for operating through a safety barrier, Fig. 4-18.

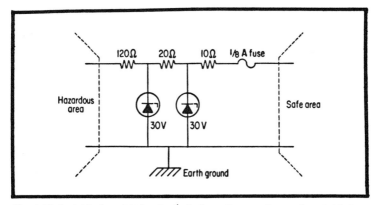

Fig. 4-18. An intrinsic safety barrier limits the voltage and current available to a circuit in a hazardous area. Shunt diodes and current-limiting resistors combine to allow an electrical signal to pass in either direction without attenuation but limit excessive current or voltage in the hazardous area.

The transmission gate, Fig. 4-19, is another useful CMOS device. When the control signal is high, transistors $Q1$ and $Q2$ are both on and may conduct current from input to output in either direction. When the control signal is low, no current can flow. This device was originally intended for use as a digital multiplexer; however, it has come to be used as an excellent analog multiplexer. CMOS multiplexers are constructed by connecting the outputs of several transmission gates.

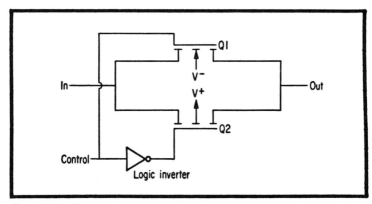

Fig. 4-19. CMOS transmission gates allow current to flow in either direction when the control signal is high; no current flows when the control signal is low. CMOS multiplexers are constructed by connecting the outputs of several transmission gates.

A CMOS transmission gate has three advantages over a relay-type gate: It has no thermoelectric effect; it has no power-consuming coil; and it has a switching speed three to four orders of magnitude faster than a relay. Its disadvantages are that it may have a resistance of several hundred ohms, and it has a lower voltage range. The resistance causes no error when the output is connected to a high impedance instrumentation amplifier; the limited voltage range, usually ±5 V, can be tolerated by most transducers such as thermocouples, RTDs, and pressure sensors.

CMOS MULTIPLEXERS

The limited voltage range of a CMOS multiplexer makes it unsuitable for a large control room application in which a multiplexer must handle sources whose common mode voltages may vary by tens or even hundreds of volts. This situation can be avoided if the entire system is configured with small remote multiplexers of 16 to 128 channels, each isolated from the control room ground by transformers and optical couplers.

If transformer or optical isolation is used, sensor signals must be chopped or digitized at the remote site. Low level analog signals are subject to noise even when chopped or converted to DC. In addition to noise, other error sources may arise, particularly when safety barriers are used (Ref. 4.1). For these reasons, an analog-to-digital (A/D) converter should be used at the remote multiplexer. CMOS circuitry can be used in the A/D for analog switching and low power logic.

When a CMOS multiplexer is used with an instrumentation amplifier, leakage current becomes the primary error source. For CMOS multiplexers, the maximum specified leakage current at a maximum temperature of 85 °C typically is 6 nA per channel. Assuming a worst case condition of one channel on and 30 channels off, a current of 180 nA would flow through a typical sensor resistance of 100 Ω, causing an 18 μV error. This corresponds to approximately 1 °C for most thermocouples under worst case conditions.

Typical measured leakage currents are less than 2 nA for an 8-channel multiplexer at an ambient temperature of 85 °C. This data was taken from several shipments of the most inexpensive type of CMOS multiplexer from two different manufacturers. With careful selection, leakage current values of less than 0.2 nA can be obtained.

Inexpensive CMOS multiplexers may require special precautions to prevent damage from transient overvoltages or power-up surges. Both the supply current and input current should be limited, Fig. 4-20, to prevent latch-up conditions caused by parasitic substrate transistors. Several manufacturers of CMOS circuits have produced special designs that prevent latch-up. These designs (Ref. 4.2) are known as

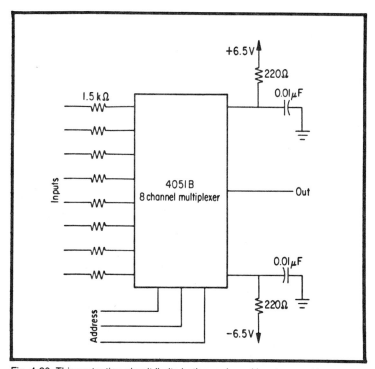

Fig. 4-20. This protection circuit limits both supply and input current to prevent latch-up during overvoltage or power-up conditions.

buried layer, floating body and dielectric isolation. However, these devices are considerably more expensive than ordinary CMOS multiplexers, typically costing about one dollar per channel. The 4051B, 4052B and 4053B family of CMOS multiplexers, available from several manufacturers, is presently selling for less than 15 cents per channel in volume.

SAFE AMPLIFIERS

CMOS devices serve other useful purposes in remote data acquisition equipment. Sensor voltages, for example, must be amplified prior to being processed by an A/D converter. A high performance instrumentation amplifier can be constructed from micro-powered integrated circuit operational amplifiers. Such an amplifier will draw less than 100 μA from its power supply and have very low input bias currents, typically less than 100 nA. Input offset drifts are less than 10 μV/°C.

The last specification, offset drift, traditionally has required that matched transistor pairs or some type of temperature compensation be used. However, with the availability of low cost, low power analog switches, a system can be constructed with a simple auto zero circuit, Fig. 4-21 which will eliminate all concern about voltage offset vs temperature. Prior to each input scan, the multiplexer is addressed to a shorted input; when the integrator switch is turned on, the amplifier output quickly will go to zero. When the switch is off, the integrator capacitor will maintain an error correction voltage.

In addition to auto zero, CMOS switches can also be used for an auto ranging amplifier with no penalty in terms of power consumption or accuracy. Accuracy is maintained by having only the operational amplifier bias currents flow through switches, eliminating concern over the high "on" resistance of each switch. Gain switching makes it feasible to provide a reference diode at the remote site which can be sampled once

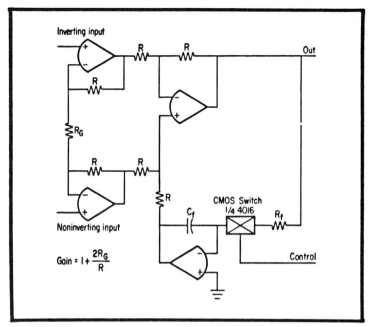

Fig. 4-21. By addressing a shorted input before each scan, an auto zero function can be added to an instrumentation amplifier. Auto ranging can be implemented by addressing a reference diode.

per scan for automatic gain calibration. Most reference diodes require 7.5 mA at 6.2 V, a large part of the power budget for an intrinsically safe circuit. Reference diodes are now available, however, with 10 ppm/°C drift and operating currents of 100 μA.

LOW POWER A/D

With CMOS circuits, micro-powered operational amplifiers, and low current reference diodes, all the essential devices are available to build a low power remote A/D converter. Such equipment has already been built with so much accuracy that virtually the only source of error is the signal-to-noise ratio in the data transmission path.

An instrument designer can't anticipate the signal-to-noise ratio present at each installation; as a result, it is best to construct an A/D which can be used with variable data rates.

For an ideal binary transmitter, the data rate can be calculated (Ref. 4.3) with the following equation:

$$C = B \ln (1 + S/N)$$

where

C = data rate in bits/s
B = bandwidth
S/N = signal-to-noise ratio

A convenient scheme for building a variable rate A/D has been developed using a voltage-to-period converter with a carrier frequency of 13 kHz. In low noise or low resolution applications, every period may be interrogated by the receiver. Allowing for settling time and synchronization, a channel throughput rate of 3 to 5 kHz is possible. In other situations, accuracy specifications or noise may require that the receiver average some convenient number of periods. A channel throughput rate of 30 to 300 Hz is more typical in these instances.

It is possible to build a complete A/D converter which will produce binary, BCD or ASCII data at the remote site. This data can then be transmitted in parallel or serial. Although binary coding is the most bandwidth-efficient method of data transmission, the random noise bursts and electrical pulses of industrial environments can degrade the signal rapidly.

A noise burst during transmission which changes the most significant bit of a data value causes a 50 percent error. With a voltage-to-frequency or voltage-to-period converter, the effect of a noise burst is at least more predictable and can be reduced to affecting only a few of the least significant bits. In addition, a voltage-to-period converter is an integrating, dual-ramp circuit which has good input noise rejection.

REMOTE DATA ACQUISITION

A system has been developed at Burr-Brown Research Corp. which incorporates some of the circuits discussed in this article. The Micromux, Fig. 4-22, is a 16-channel remote I/O

Fig. 4-22. The Micromux incorporates CMOS logic for remote data acquisition. Each remote multiplexer unit, equipped with an A/D converter and a serial interface, can handle up to 16 analog and digital channels.

multiplexer which can accept thermocouples, 4–20 mA signals, voltage signals and contact closures. An integrating voltage-to-frequency device performs A/J conversions; digital outputs are transmitted over a twisted pair to a receiving module. Data is transmitted in a current loop to avoid noise. Signals can be transmitted up to 1,500 m (5,000 ft).

The same twisted pair used to transmit data also supplies all the necessary power for the CMOS circuitry so that the multiplexer requires no local power at its point of installation. The receiver, designed to interface to most common minicomputers, can handle up to four remote multiplexers; up to eight receivers can be connected to one computer interface for a total capacity of 512 channels.

INTELLIGENT SENSORS

The principal problem with using CMOS multiplexers in remote I/O applications is their limited channel-to-channel common mode voltage range. The semiconductor industry, however, is rapidly making an economic reality of an old idea which would solve this and many other problems: an A/D converter at every point.

In Fig. 4-23A, a monolithic instrument amplifier, A/D and UART are combined with some control circuitry and an optical coupler to provide a module which converts a transducer signal to a digitized serial ASCII output. Power for the circuit can come from an external source or it can be carried on the twisted pair used for communication. The multiplexer in both figures is optional.

A microprocessor and PROM perform the A/D and UART functions in Fig. 4-23B. Since most microprocessors are run from a crystal-controlled oscillator, they have a highy accurate timing capability. When an accurate time base is combined with a precision voltage or current source under the control of a microprocessor, several simple A/D conversion circuits are possible. A crystal oscillator-based voltage-to-frequency converter is one possibility. The simplest is a precision pulse width modulator and analog comparator. Figure 4-23B shows a monolithic D/A and analog comparator which combine with the microprocessor to form a high speed successive approximation A/D.

The microprocessor-based device offers a unique control system advantage: at almost no increase in cost, it could be programmed for local digital control. Through its serial ASCII link to a supervisory computer, it could accept setpoints and other control information and effectively implement almost the ultimate in total distributed control. Using discrete chips, it is possible to build such devices today for under $150 per point; in five years, these devices will be available in highly integrated form and sell at a much lower price.

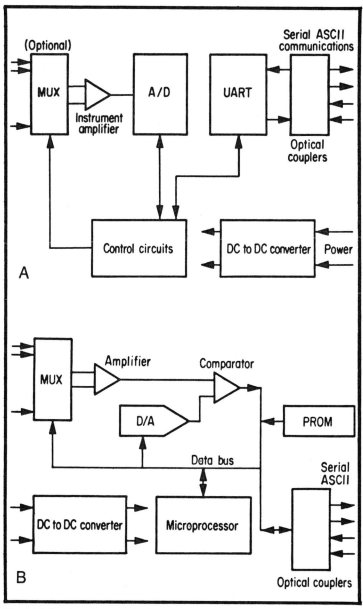

Fig. 4-23. Semiconductor technology is making the old idea of an A/D at every point an economic reality. A) In this circuit a monolithic A/D digitizes a signal and a UART transmits it in serial ASCII to a supervisory system. B) A microprocessor replaces the A/D and UART in this circuit. A/D conversions are performed with a monolithic D/A and an analog comparator in a successive approximation technique. The microprocessor can also be used for control purposes. Both circuits can be built with currently available chips for under $150.

REFERENCES

4.1 Tiffany, F. and Pederson, R., "Barriers Can Alter Circuit Parameters," *Instruments & Control Systems*, August 1975, pp. 25-28.

4.2 Thibodeaux, E., "Getting the Most Out of C-MOS Devices for Analog Switching Jobs," *Electronics,* December 25, 1975, pp. 69-74.

4.3 Shannon, C.E., "Communications in the Presence of Noise," *Proceedings of the Institute of Radio Engineers*, Vol. 37, January 1949, pp. 10-21.

LEE PAYNE is a design engineer at Burr-Brown Research Corp., Tucson, AZ. The article is based on a paper presented at the ISA Computer Interface Instrumentation Symposium, Newark, DE, 1976.

Common-Mode Rejection Techniques for Low-Level Data Acquisition

M. B. COFFEE

Prospective users of computer-controlled data acquisition systems have to choose not only a manufacturer but very often a technique for acquiring low-level data. Techniques used in achieving common-mode rejection can have a decisive influence on the cost, ease of implementation and final performance of an analog input system.

Analog voltage measurement, in nearly all computer-controlled, low-level data acquisition systems, is differential. This means that each data signal is the voltage difference between two input points. A voltage which exists in each of the differential inputs is a *common-mode voltage*.

Common-mode rejection (CMR) techniques discussed in this article represent different methods for preventing the common-mode voltage (CMV) from being converted into a normal mode or differential voltage, and appearing as a data signal at the output of the circuit's differential amplifier.

In many cases, CMV is a product of the circuit producing the differential voltage, as in the case when measuring voltage across an ungrounded resistor. However, the most common situation is when CMV simply is the result of a difference in potential between two physically remote electronic grounds.

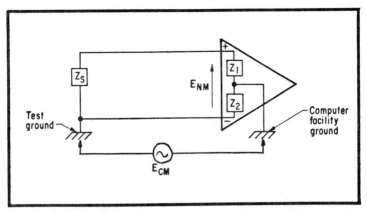

Fig. 4-24. Common-mode voltage E_{CM} represents the difference in potential between two physically remote grounds. Techniques used to achieve common-mode rejection are designed to prevent the common-mode voltage from becoming a data signal.

This situation always exists, but the pertinent question is whether or not the voltage is significant when compared to the signal being measured. In an industrial plant, this voltage often is on the order of 1 to 2 V, even when considerable care is taken in grounding the computer system and the circuit being measured.

Differences in potentials are due to the fact that most electronic equipment and power distribution circuits are connected to earth ground for safety or convenience, and large currents result. Particularly bothersome with some types of analog systems is the high-frequency CMV produced by nearby radio stations, radar installations, induction heaters, motor control relays and other sources of electrostatic and electromagnetic noise.

A common-mode voltage error will result if any of voltage E_{CM} is converted to a normal-mode voltage E_{NM}, Fig. 4-24. The common-mode rejection ratio (CMRR) is defined as:

$$CMRR_{db} = 20 \ log_{10} \frac{E_{CM}}{E_{NM}}$$

A more detailed representation of differential input cabling, Fig. 4-25, shows a situation that can occur when input leads are unshielded, and current flows in the input leads because of the common-mode generators. Common-mode errors occur when the stray capacitance and distributed resistances are unbalanced ($R_s \neq 0$), which is almost always the case. Perfectly balanced input lines simply don't happen. Sometimes attempts are made to reduce the common-mode voltage by installing heavy bus bars and braid between the source ground and the computer facility ground.

COMMON TO NORMAL-MODE CONVERSION

Most common- to normal-mode conversion is caused by currents which flow in the input leads when the impedance in each lead is not perfectly balanced. Referring to Fig. 4-24, if Z_1 and Z_2 are equal to Z_{IN}:

$$E_{NM} = \frac{E_{CM}\, Z_{IN}}{Z_s + Z_{IN}} \quad -E_{CM} = \frac{E_{CM}\, Z_s}{Z_s + Z_{IN}}$$

Voltage E_{NM} will decrease as Z_{IN} increases and Z_s decreases. It is therefore meaningful to specify CMRR while noting the amount of source unbalance present; most manufacturers specify CMRR at $1,000\ \Omega$ source unbalance. Also, the CMV

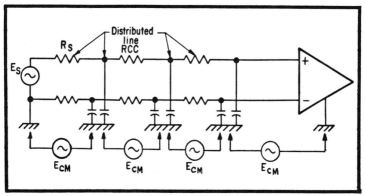

Fig. 4-25. When a number of physically remote grounds are present in a system, common-mode errors occur because of stray capacitances and unbalanced resistances in the circuits. This can be reduced sometimes, by installing bus bars or braid between the source grounds and the computer facility ground.

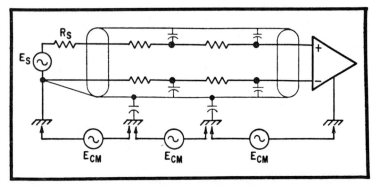

Fig. 4-26. A guard shield, connected at the signal source and capacitively coupled to ground, provides a conductive path for the common-mode current. Shield effectiveness depends upon its coverage of the circuit and its conductance.

typically is not DC; therefore, the reactive component of the input coupling to ground is particularly important. Because of this, and the fact that the CMR of the front end varies with frequency, most manufacturers quote CMRR within certain frequency limits. A typical spec would be "CMRR = 120 dB, 0 to 60 Hz with 1,000 Ω source unbalance." This spec represents an impedance to ground of 1,000 mΩ. If the impedance is entirely capacitive, it represents a capacitance of only 2.7 pF at 60 Hz.

One of the most practical approaches to minimizing common- to normal-mode conversion is to reduce the reactive leakage currents flowing from the signal leads to ground. The most frequently used method is the *guard shield*, a conductive path between the signal leads and ground, Fig. 4-26.

When a guard shield is maintained at the common-mode voltage, induced currents in the signal leads are held to a minimum. The shield is capacitively coupled to ground, allowing common-mode induced currents to flow in the shield, rather than in signal leads. For this reason, shield effectiveness is increased as its conductance (and of course its coverage) is increased. For a shield to be most effective as a common-mode guard, it should be referenced to the CMV at the transducer, as shown in Fig. 4-26.

Five guard shield and amplifier input configurations are used in most of the low-level channels being installed today:

- amplifier per channel
- floating front end
- active guard
- flux coupled
- flying capacitor.

Any of the five techniques, if properly implemented, will yield CMRRs of 120 dB, 0 to 60 Hz at 1,000 Ω source unbalance, or better. The technique chosen in a low-level data acquisition system is affected most significantly by the channel sample rate and the maximum common-mode voltage required.

AMPLIFIER PER CHANNEL

With this technique, a differential amplifier is used with every transducer, Fig. 4-27. This approach allows high system throughput (up to 70,000 samples/s) because high-level, solid-state multiplexing may be used, The amplifier usually is solid state and direct-coupled, and yields a signal plus common-mode capability of 12–15 V.

Because each channel has an amplifier, the cost per channel of this configuration is increased; but this cost is mitigated to a degree because the multiplexer need only

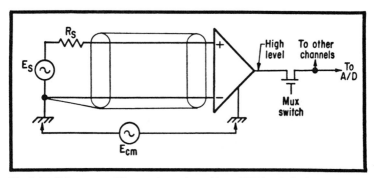

Fig. 4-27. Using an amplifier per channel to achieve CMR is relatively expensive, but it permits rates of up to 70,000 samples/s because a high-level, solid-state multiplexer can be used.

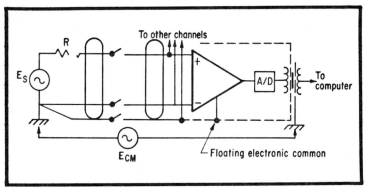

Fig. 4-28. A floating front end "assumes" the CMV immediately before the multiplexer samples the data point. Speed is sacrificed because of the relay switching involved, but the system can handle CMV's to ±200 V.

switch single-ended, high-level signals. Systems requiring wide dynamic range often are implemented with an input system that has programmable (and sometimes autoranging) amplifier gains. Wide dynamic range is expensive to implement on a per-channel basis and even more so if the amplifier is isolated. Amplifier-per-channel systems have relatively high maintenance costs since each channel requires periodic "tweaks."

FRONT END FLOATING

The floating front end technique, Fig. 4-28 can have a high common-mode voltage capability (±200 V), since the amplifier is not directly connected to the computer ground; instead, it "floats" on the common-mode potential.

Speed is sacrificed to obtain the higher CMV capability. Because of the higher CMV and lower speed, this approach uses relay switching and an integrating A/D converter. The guard relay is closed before the signal relays close to allow the shielded front end to "assume" the CMV. Integrating (dual slope) or voltage-to-frequency type A/D converters often are chosen because their speeds are compatible with relays, they are isolated without a large cost, and both provide noise rejection through their integrating action.

ACTIVE GUARD

The active guard technique, Fig. 4-29, is used in high-speed (20 kHz) systems with direct coupled MOS-FET multiplexers. High CMV capability and CMR at higher frequencies are sacrificed to obtain the faster speed.

A wide-band amplifier reflects lower common-mode reactance into individual channel shields, thereby increasing the speed. The trade-off is that the guard driver will react to high-frequency noise which can cause it to saturate and give erroneous results to subsequent channels.

HIGHER CMV

Flux-coupled or transformer-isolated multiplexers use transformers to obtain common-mode isolation and rejection, Fig. 4-30. The resulting common-mode capability is up to 2,000 V with system throughput reaching 20,000 samples/second.

Chipper signals modulate the input signal, allowing it to be coupled to the transformer secondary. The resulting AC signal is then amplified, demodulated and coded. This technique eliminates the need to switch the shield since it is always at the common-mode voltage. The high gain amplifier, since it is AC-coupled, does not contribute to the DC drift. This technique maximizes gain and noise performance.

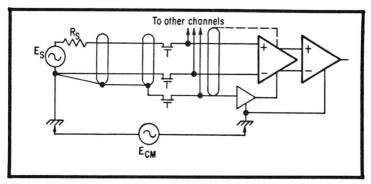

Fig. 4-29. The active guard technique uses a wide-band amplifier to "reflect" a lower common-mode reactance into the shield input, increasing speed.

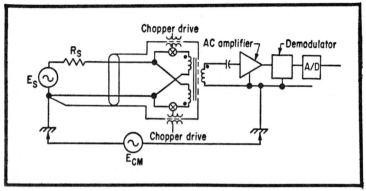

Fig. 4-30. Transformers and a chopper can be used to transformer isolate the DC signal being measured. The shield does not have to be switched because it is always at the CMV. This system can handle CMVs up to 2,000 V.

This scheme, because of the precision transformers on each channel, generally is higher in cost per channel than other techniques. Benefits include extremely high common-mode voltage capability, high sample rates and electronic reliability.

CAPACITOR CHARGING

A flying capacitor front end, Fig. 4-31, removes the common-mode voltage from the signal by charging a capacitor to the differential voltage and then disconnecting the capacitor from the input circuit and connecting it to the input amplifier. The flying capacitor is capable of handling large common-mode voltages since the switching device need only withstand the CMV of the channel being sampled (the low side of the common analog bus is connected to the ground).

Systems which use differential analog buses must have switches capable of handling twice the CMV since a switch on an unselected channel will have a voltage across it equal to the algebraic sum of its own CMV and the CMV switched to the bus from the selected channel. The front end amplifier used on a flying capacitor system need not be differential; this results in a simpler amplifier.

Disadvantages of this approach are that each channel must have a double pole, double throw switch. The flying cap does not provide the integrating A/D conversion and, therefore its low frequency noise rejection, but does provide excellent immunity from higher frequency CMV.

HOOKING UP

In all techniques, proper installation of field wiring is of extreme importance when a low-level signal must be extracted from a significant common-mode voltage. Although it is not always possible to do so, the signal leads should be shielded the entire distance between the signal source and the analog system input connectors. Foil shields usually are preferred to the braided variety because foil offers 100 percent coverage and ease of handling. The remaining question is where to connect the shield for a given situation.

Thermocouples frequently are bothersome because they may start out grounded and become ungrounded (or vice versa), or may be intermittant. Grounded, shielded thermocuples represent the best situation. Ungrounded thermocouples should have their shields grounded and connected to the negative lead at a convenient point. This is usually done at a connection panel located close to the probe or the computer system.

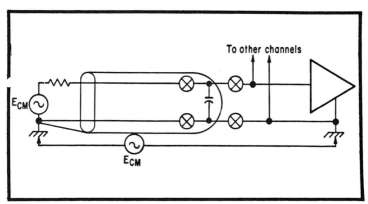

Fig. 4-31. A flying capacitor front end removes CMV by charging a capacitor with the data signal and then connecting the capacitor to the input amplifier. Large common-mode voltages are handled because switching devices need only withstand the CMV of the channel being sampled.

Bridge-type inputs are either grounded at the transducer or at the excitation supply. If the transducer is shielded, it may be grounded locally with the shield connected to the low side of the excitation supply at the transducer. Otherwise, the low side of the excitation supply should be grounded; this point should also serve as the shield ground.

Very often a computer-based data acquisition system is installed as a replacement for a stripchart system, and existing wiring is used. Since the bandpass of a stripchart system is very low, the guard shield techniques explained here probably were not employed. When old wiring is used, it does no good to specify high common-mode rejection for the analog input subsystem since the common-to normal-mode conversion takes place in the input cabling.

Concern with the type of system, its quality and the techniques used in the input wiring will have a rewarding payoff as soon as the system is installed. As in so many other aspects of designing a data acquisition system, wrong choices are very difficult to reverse after the system is in place.

MICHAEL B. COFFEE is Midwest Regional Sales Manager, Computer Products, Inc., Ft. Lauderdale, FL.

Accuracy In Signal Conversion

D. W. DEVALL

In precision measurement involving highly accurate primary transducers, system accuracy may be limited by the signal converter. This problem is frequently encountered with RTDs, strain gauges, and measuring systems based on precise resistance or low-level DC signals.

A signal converter is a device which produces an electrical output compatible with a receiving element in response to an input signal or condition. In most industrial instruments, the output is a current signal; the input may be millivolts, resistance, pressure, or other quantities. The converter performs two functions:

- Transforms the input quantity into an electrical signal, if necessary
- Amplifies the electrical signal to the desired output level.

In a resistance-to-current converter, Fig. 4-32, the primary transducer is an RTD, which is connected in a balanced bridge circuit to produce millivolt signal proportional to its resistance. The millivolt signal is then raised to the required output level by the amplifier A.

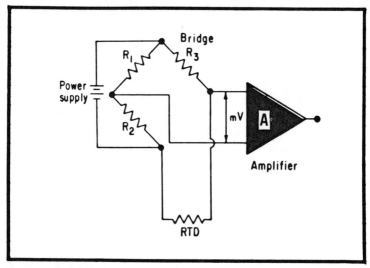

Fig. 4-32. Resistance-to-current converter.

SOURCES OF ERROR

In general, errors can be introduced in either the conversion section or the amplification section or both. Referring to Fig. 4-32, errors in the conversion section may be due to: 1) variations in the bridge excitation voltage E; 2) variations in the resistances R_1, R_2 and R_3 because of temperature changes; and 3) poor connections to the RTD, noise pickup to thermal EMF effects on the input lines.

Amplifier section errors may result from nonlinearity, hysteresis, ambient temperature effects, vibration, power supply effects, drift and common mode interference, among other factors. Nonlinearity, hysteresis and drift characteristics are inherent to the instrument; while any errors due to ambient temperature variations, vibration, common mode and power supply effects depend upon the particular installation.

Most commercial signal conversion devices incorporate designs to minimize these errors. Such designs include a matched differential input amplifier, low-temperature-coefficient components in critical circuits, regulated power supply, and possibly temperature compensation. Accuracies of commercially available signal converters range from ±0.25 to ±0.1 percent, with typical stability of ±0.5 percent per month.

In the selection of an instrument to be used for high precision measurements, the accuracy, reproducibility, and stability specifications are general guidelines as to the instrument's performance. But, seldom are the causes of error and the relative magnitudes of errors due to each cause specified. When problems arise due to such unknown effects, an instrumentation amplifier can offer an effective solution by providing a stable, reproducible, and predictable output signal.

INSTRUMENTATION AMPLIFIERS

The instrumentation amplifier is basically a differential input, closed-loop, variable-gain operational amplifier. The primary sources of error attributable to the amplifier are temperature drift and the effects of power supply variations. Hysteresis and nonlinearity are negligible for a good quality instrumentation amplifier, whose typical specifications are shown in Table 4-5.

The system shown in Fig. 4-33 is a millivolt-to-current converter built around a typical instrumentation amplifier. Input voltage V_t is amplified to provide output current I_o at the corresponding voltage level V_o. The receiver input impedance is R_r, and R_a is a current-limiting resistor.

Table 4-5. Instrumentation Amplifier Specifications

Gain	1.0 to 1,000	
Temperature drift	±2 μV/°C	(RTI)[1]
Drift stability	±10 mV	(RTO)[2]
Power supply effect	±20 μV/%	(RTI)[1]
Input impedance	10 MΩ	
Common mode rejection	120 dB	
Output voltage, maximum	±10 V @ 5.0 mA	
Output resistance	2.0 Ω	
Linearity	0.02%	

Note: 1—Referred to input
2—Referred to output

393

One must be sure that the output current specification is not exceeded. The value of R_a is determined by the maximum values of output voltage and current, with allowance for the current-limiting effect of the receiver input impedance:

$$R_a = \frac{(V_o)_{max}}{(I_o)_{max}} - R_r \qquad (4.1)$$

The amplifier gain A is the ratio of output voltage span to input voltage span:

$$A = \frac{(V_o)_{max} - (V_o)_{min}}{(V_i)_{max} - (V_i)_{min}} \qquad (4.2)$$

The zero (offset) and span (gain) are each adjustable by means of a single control.

CALCULATING PERFORMANCE RTO

An application example, based upon Fig. 4-33 and the specs in the Table, shows how to calculate the performance of the conversion system. The following input and output ranges are assumed:

$$V_i = 0 \text{ to } 20 \text{ mV}$$
$$I_o = 1 \text{ to } 5 \text{ mA}$$
$$V_o = 2 \text{ to } 10 \text{ V}$$
$$R_r = 50 \text{ } \Omega$$

In this case, from Equations 4.1 and 4.2,

$$R_a = \frac{10}{0.005} - 50 = 1,950 \text{ } \Omega$$

$$A = \frac{8}{.02} = 400$$

The specifications in the Table state temperature and power supply drift effects in terms of input signal ranges

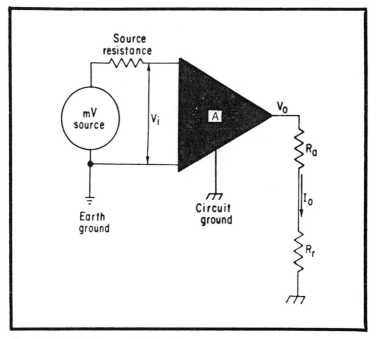

Fig. 4-33. Millivolt-to-current converter.

(RTI). To evaluate system performance, these effects must be calculated in terms of the converter-amplifier output (RTO). The temperature drift (TD) referred to the output is:

$$TD_{RTO} = (TD)_{RTI} \times A \qquad (4.3)$$

$$= \pm 2\ \mu V/°C \times 400 = \pm 0.8\ mV/°C$$

The power supply effect (PD) referred to the output is:

$$(PD)_{RTO} = (PD)_{RTI} \times A \qquad (4.4)$$

$$= \pm 20\ \mu V/\% \times 400 = \pm 8\ mV/\%$$

These drift figures can be expressesd as a percent of fullscale output, if divided by $(V_o)_{max}$:

$$TD = \pm 0.8\ mV/°C \div 10\ V = \pm 0.008\%/°C \quad (4.5)$$

$$PD = \pm 8\ mV/\% \div 10V = \pm 0.008\%/\% \qquad (4.6)$$

Equations 4.5 and 4.6 express the output signal variation, in percent of fullscale, for a one-degree drift in temperature and for a one-percent drift in power regulation.

If the power supply used is regulated to ±0.1 percent, the power supply effect will be ±.008 percent of the fullscale output—or ±0.8 mV. From the spec table, long-term stability is ±10 mV at the output or ±0.1 percent. Assuming that the errors (TD and PD) are additive, this accuracy of ±0.1 percent can be achieved over a temperature range of ±11.5 °C:

$$\frac{0.1\% - 0.008\%}{0.008\%/°C} = 11.5 \ °C$$

Note that the drift figures are directly related to the amplifier gain. In the above example, if the input span were 0 to 50 mV, accuracy of 0.1 percent could be maintained over a temperature range of ±30 °C.

Adapting this amplifier to a resistance bulb input signal involves a bridge circuit, Fig. 4-34. The resistances R_1, R_2 and R_3 are each equal to the nominal resistance of the RTD at its operating midpoint. R_4 is much smaller in value and is used to adjust zero. Span of the bridge output signal is determined by the bridge excitation voltage B; and is adjusted by varying the amplifier gain. A well-regulated power supply and low-

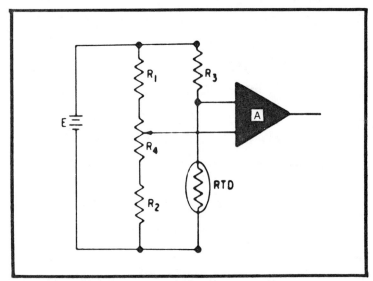

Fig. 4-34. RTD bridge circuit.

temperature-coefficient resistors are necessary for good accuracy. The same principles apply to strain gauge inputs, thermistors, and other resistance signals.

Specifications for instrumentation amplifiers vary; output current ratings up to 20 mA are available, as are lower temperature coefficient and drift specifications than those shown in the Table. The application dictates the exact requirements.

MINIMIZING ERRORS

There are several points to remember which will insure that signal conversion equipment performs within specification:

- Proper wiring practice, including shielding, grounding and routing must be observed. The system should be earth-grounded at one point only, preferably the signal origin. Twisted-pair shielded cable will minimize extraneous pickup from power and other circuits. Input lines should be balanced as well as possible and routed as far away from power cables as practical. Wire of similar material should be used at all junctions to avoid thermally induced EMF.

- In some cases, the choice of AC or DC power input is optional. Selection of DC input and use of a well-regulated DC power supply will reduce errors induced by the power supply to a minimum.

- Amplifier input impedance should be as high as possible, at least 1,000 times that of the signal source—especially when dealing with millivolt signals.

- Whenever possible, instruments should be placed in a temperature-controlled environment (such as a control room) to reduce thermally induced drift.

- In environments with excessive vibration, shock-proof mountings can reduce the effect on the instrument from both accuracy and maintenance standpoints.

DOUGLAS W. DEVALL is an instrument engineer assigned as Methods Engineer at Uniroyal's Painesville, Ohio facility.

Adapting Electric Actuators to Digital Control

M.F. HORDESKI

As digital control becomes more widespread in the process industry, the problem of valve control looms as a large obstacle to progress in many plants. Most existing plants have a large number of valves installed, and most of these valves have many years of useful life remaining. Also, viable maintenance policies, which include regular service and periodic rebuilding, extend valve life even further.

These factors have slowed the changeover to digital actuator valve systems. However, methods are now available to convert most electric actuators to digital control.

CONVERTING INPUTS

If an electric valve actuator is to be used with a digital control system, some means must be found to allow the digital system to control the actuator prime mover. The digital command to the valve must be converted to a form which the actuator can accept, and the control signal must be matched to the actuator's characteristic.

One method for matching the digital command to the actuator input involves the use of a digital-to-analog converter

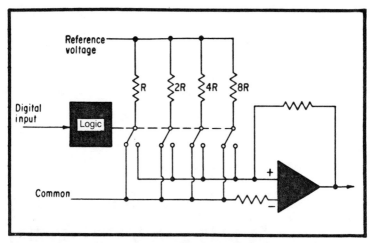

Fig. 4-35. In a digital-to-analog converter, the digital signal is switched to a binary-weighted network that supplies a voltage or current signal proportional to the value of the digital signal. This figure shows a DAC that outputs a voltage signal.

(DAC). A DAC consists of a binary-weighted network whose outputs are summed to produce an analog signal that is proportional to the binary or digital input, Fig. 4-35. Inputs to the weighting network are controlled by solid-state switches that convert the digital signal to the proper binary form; the output of a DAC can be either current or voltage.

A typical electronic DAC consists of solid-state switches, logic to control the switches, and either a resistive ladder network for voltage output or a set of current sources for current output. A DAC can be built by a user with available components, but this is done only on rare occasions, such as in specialized, extremely high-speed applications. Errors in matching off-the-shelf components can quickly add up, sometimes obliterating the D/A conversion process. It is much easier to purchase DACs from a manufacturer. Electronic DACs are available as monolithic chips, chip sets or potted modules at inexpensive prices.

To minimize noise and cabling problems, the DAC should be located as close to the actuator as is possible. The actuator's electrical load should be matched or filtered to prevent the actuator from responding to spikes caused by switching time differences in the DAC's weighting network.

MATCHING AND SCALING

Matching a DAC output to the electrical load of a high-current electric actuator can be accomplished by using the actuator load as a feedback element to an operational amplifier, Fig. 4-36. The operational amplifier is controlled by the DAC output voltage. Feedback resistance r_f is adjusted to equal the DAC output voltage divided by the desired current:

$$r_f = \frac{e_o}{i_L}$$

Capacitor C1 and resistor R1 can be used to reduce oscillations caused by parasitic impedance in the load and wiring, and a clamping diode (D) can be placed across the coil to prevent high voltage transients from appearing at the operational amplifier output. This diode also can improve time response by allowing load current to decay quickly. The circuit in Figure 4-36 can be used to control dc-operated directional valves on electrohydraulic actuators, or gear-driven actuators controlled by dc motors.

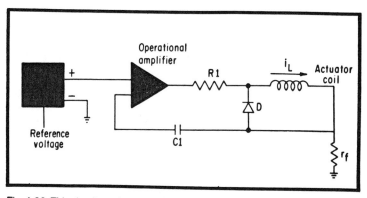

Fig. 4-36. This circuit can be used to control DC-operated directional valves and actuators that use DC motors. To match the DAC output to the high-current load of the actuator, the actuator load is used to provide feedback to an operational amplifier.

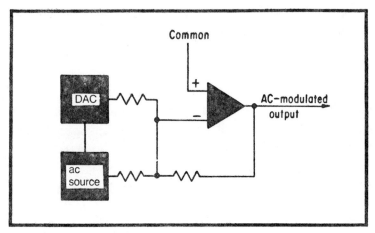

Fig. 4-37. A multiplying DAC produces, in this case, an AC-modulated signal that can be used to control AC-driven directional valves and AC motors. The AC source supplies a dither signal to prevent silting contamination in spool valves.

Controlling ac loads involves the use of a multiplying DAC. Standard DACs require a reference voltage that varies only a small amount, usually within ±5 percent of the fullscale operating voltage. A multiplying DAC, however, permits the reference voltage to vary widely since the output is obtained by internally multiplying the reference voltage by the digital input.

In Fig. 4-37, the reference voltage for the multiplying DAC is also connected to an operational amplifier. The reference is permitted to vary in accordance with the AC load frequency; the output, therefore, is an AC-modulated signal. The percent modulation is dependent upon the digital input signal to the DAC. This type of AC modulation can be used to control AC-driven directional valves and AC motors.

Many control valve applications will require both an AC scaling and a multiplying DAC, and an output amplifier circuit such as the one shown in Fig. 4-36. AC scaling must be used in some cases to reduce voltage to a level that the multiplying DAC can accept. Also high AC voltages require that amplifiers be constructed of discrete semiconductor devices, similar to those used in motor control modules.

Commercially-available motor controllers employ SCRs or triacs to allow the motor reversal required for gear-driven actuators. Many installations include a motor controller as part of the actuator system and it is necessary to provide only a matched or scaled analog signal from a DAC to convert them to digital control.

MONITORING POSITIONS

Many methods are available to monitor the position of a valve stem or gear shaft. LVDTs, limit switches and linear potentiometers have been used on stem actuators, while rotary potentiometers and limit switches have been used on quarter-turn actuators for ball and butterfly valves.

Limit switches are used primarily to provide "open" and "closed" information, with a few intermediate positions available in some cases, but limit switches do not provide the degree of information required for accurate control. LVDTs and potentiometers both require conversion electronics to provide an output signal in digital form. Synchros, which basically are variable rotary transformers, can be used to indicate rotary position on ball and butterfly valves. Synchros are not normally used, however, because of cost and complexity.

Digital encoders are well suited for valve position feedback: both linear and rotary types exist, no conversion electronics are required, and the device can be attached to an actuator as easily as can rotary or linear potentiometers. Several types of digital encoders are available:

Conducting encoders use brushes or wipers to detect the position of a coded disc or plate, Fig. 4-38A. The coded portion is plated with a precious metal alloy for good conduction, and the wipers are made from the same material to minimize arcing and material migration. Even with these features, which drive up the cost, the conducting encoder has a limited life and should be used only where the application permits easy replacement and downtime is not critical.

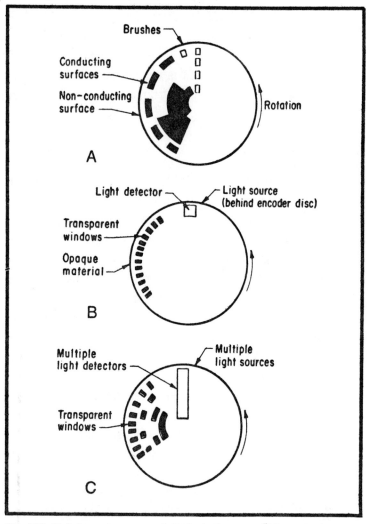

Fig. 4-38. Digital encoders are available for both rotary and linear applications. These rotary encoders employ various methods of determining position: A) A conducting encoder detects the presence of coded conducting material to calculate the absolute position of the valve. B) An incremental encoder outputs a pulse when the light detector encounters a transparent "window." The control system has to keep track of valve position by adding or subtracting pulses to the current position count. C) An absolute optical encoder has multiple tracks and detectors to determine valve position.

Optical encoders have a coded member (disc or track) with light-transparent "windows" that represent a digital code. A light source is placed on one side of the coded member and light-detecting semiconductors are placed on the other side. If a single track of transparent windows is used, Fig. 4-38B, the encoder outputs a pulse each time a space is detected. This type of device is called an *incremental encoder*. To determine valve position, the control system must count each pulse and add (or subtract) it to the current calculated position. Multiple tracks and detectors can be used to indicate absolute position, Fig. 4-38C; these devices are called *absolute optical encoders*.

ENCODER CODING

Multiple-track encoders can have erroneous outputs at intermediate positions unless precautions are taken. Table 4-6 shows the three most popular codes in use today: binary, binary coded decimal (BCD), and Gray or cyclic binary. The first two codes, binary and BCD, allow more than one bit to change at intermediate positions; e.g., from 001 to 100,

Table 4-6. Digital Codes

Binary

Decimal value	8-bit output
0	00000000
1	00000001
2	00000010
3	00000011
4	00000100

Binary coded decimal (BCD)

Decimal value	8-bit output (2 digits)
0	0000 0000
1	0000 0001
2	0000 0010
3	0000 0011
4	0000 0100

Cyclic binary

Decimal value	8-bit output
0	00000000
1	00000001
2	00000011
3	00000010
4	00000110

Table 4-7. Resolution Equivalents

Word length (bits)	Number of positions available	Resolution (percent fs)
4	16	6.25
5	32	3.12
6	64	1.56
7	128	0.78
8	256	0.39

where three bit values change. In cyclic binary, only one bit changes between positions and the value of the maximum error, therefore, is limited to the least significant bit.

Mechanical and electrical techniques are available to minimize errors at intermediate positions. For conducting devices, an extra set of brushes can be installed near the primary set. Outputs from the second set of brushes can be processed by antiambiguity logic which compares the two outputs and discards erroneous codes. In optical devices, the same function can be obtained by installing secondary light sources and detectors.

In addition to selecting an appropriate code for the position monitoring device, it is necessary to determine resolution and the type of electronic logic interface. Resolution can be calculated by dividing the amount of valve travel (linear or angular) by the number of encoder positions, Table 4-7. Resolution can also be expressed as percent fullscale error as shown in Table 4-7.

The necessary logic interface is usually determined by other digital components in the system, but in some cases it may be desirable to convert from one type to another. For example, it may be desirable to utilize CMOS components in high-noise areas. TTL and CMOS are commonly used in many industrial systems, and converting from TTL to CMOS is no problem. Converting from CMOS to TTL is more difficult because the CMOS gates must be paralleled to drive the higher current TTL gates.

Characteristics of various integrated circit logic families are shown in Table 4-8. ECL and PMOS are more likely to be encountered within a digital control system than at the interface, since ECL is often used in large, high-speed computers and PMOS is used in many LSI circuits.

FINE TUNING

Electrohydraulic actuators use an electrically operated valve to control the flow of hydraulic fluid into an actuating cylinder which, in turn, positions the valve stem. A jet pipe or spool valve is used to divert the fluid from one side of the actuating piston to the other. The electric valve—sometimes called a servovalve—can be supplied with its own hydraulic pump, thereby forming a self-contained actuator which requires only electric power.

The low-current DC control signal required for this type of actuator can be provided by a DAC and the stem position can be sensed with a linear encoder. If the actuator uses a lever arm or rack and pinion to control a ball or butterfly valve, a rotary encoder can be attached directly to the valve for more accurate position feedback.

In an analog control system, the usual tuning practice involves increasing the controller gain (to obtain maximum dynamic response) until the actuator "jitters," goes into a limit cycle, or becomes otherwise unstable; the gain is then reduced until the actuator stabilizes. If this gain adjustment cannot be done with a digital controller for one reason or another, a simple scaling potentiometer can be added at the DAC output to fine-tune valve gain.

When a spool valve is used for fluid diversion, the analog controller often supplies a "dither" signal to prevent silting of contaminants at the spool control lands. Silting will affect low-speed operation and eventually cause erratic valve performance at all speeds. If a dither signal is needed, a small AC signal can be added as was shown in Fig. 4-37.

Table 4-8. Logic Family Characteristics

	Transistor-transistor logic (TTL)	Emitter-coupled logic (ECL)	P-channel metal-oxide-silicon (PMOS)	Complementary metal-oxide-silicon (CMOS)
Power dissipation	Medium	High	Low	Very low
Noise immunity	Very good	Good	Nominal	Excellent
Supply voltage	+5 V	−5.2 V	−27, −13 V	+4.5 to +16 V
Characteristics	Many functions available. Lowest cost. Widest range of applications. Power dissipation can be a limiting factor.	Limited functions available. High-speed applications.	Limited functions available. Widely used in custom chips. Low-speed applications.	Medium amount of functions available. Low-power applications or high-noise environments.

Some spool valves have a mechanical adjustment for nulling the output in the absence of a control signal—otherwise, the actuator would tend to drift. This mechanical adjustment is made with all electrical signals removed. A null bias signal is also used to bring the valve to a stationary closed position; the null bias signal is usually summed with the input signal at some point in the control loop. In digital systems, the reference signal into the DAC can be varied or a bias signal can be applied to the operational amplifier. In both cases, a simple voltage-scaling potentiometer can be used. Null bias drift effects due to temperature or component aging are much less in a digital system than in its analog counterparts.

An encoder can be nulled by mounting the unit while the actuator and encoder are both at null positions. Some method should be employed to lock or hold the actuator shaft in place when a rotary encoder is being installed on ball or butterfly valves. For linear encoders, Fig. 4-39, a screw adjustment should be provided for fine tuning of position.

These tuning techniques also can be employed with gear-driven actuators such as quarter-turn, linear and continuous rotation actuators, and large gear-driven actuators which use slip clutches. Converting electric actuators to digital control, even in large valve configurations, provides improved

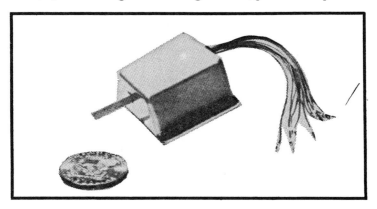

Fig. 4-39. This linear digital encoder detects the absolute position of valve stems with up to one inch of travel. A threaded control rod is used to fine-tune the device.

performance and protects and prolongs the life of these valuable components.

We wish to thank Mr. Jack Binford of Binco, Charlotte, NC, for his assistance in preparing this material.

MICHAEL F. HORDESKI is President of Siltran Digital Co., Silverado, CA.

Automatic
Bridge Balancing Circuit

J.G. STEPHENSON

A monolithic digital-to-analog converter (DAC) and a feedback loop simulate the typical manual bridge balance circuit used in strain gauge calibration. The circuit maintains system isolation and can operate with constant current or constant voltage power supplies.

Calibration and setup for a strain gauge system frequently takes more time than an actual test. Because of drifts within gauges and associated electronics, the DC balance of each gauge channel should be adjusted just before the beginning of each test cycle. Balancing a bridge is usually a manual procedure which sometimes does not take amplifier zero drift, a significant source of error, into account.

Consecutive tests may not allow enough time between tests for accurate rebalancing, especially if many channels are involved. For years, instrumentation engineers have wanted to reduce the setup and calibration times normally associated with strain gauge measurements.

MORE EFFICIENT SYSTEMS

The first time-saving bridge balance configuration incorporated a voltmeter connected to a single monitor bus. Each

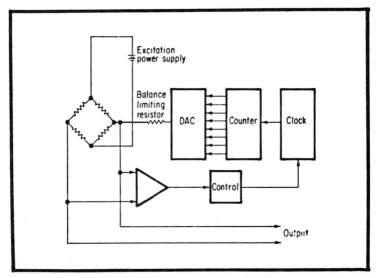

Fig. 4-40. Block diagram of typical automatic bridge balancing system. A digital-to-analog converter (DAC) provides a balancing voltage to zero the bridge during calibration.

strain gauge signal conditioner contained a bridge balance potentiometer and a switch. The switch was used to connect the bridge or amplifier output to the monitor bus and the voltmeter. This configuration at least eliminated the need to move the voltmeter from one channel to the next.

A servo-controlled potentiometer and a servo amplifier provided the first hands-off automatic bridge balance circuit. This circuit would adjust the bridge output to zero on an external command. Since the servo amplifier circuit was usually contained within the strain gauge bridge, it could not correct for amplifier zero errors. These systems were fairly expensive and had the reliability problems inherent with electromechanical systems.

Finally, the digital-to-analog converter was introduced to the problem of automatic bridge balancing. These systems varied from fully automatic hardwired circuits to computer-controlled systems.

In a DAC balancing circuit, Fig. 4-40 the unbalanced output of the strain gauge bridge drives a timer and counter which feed a digital-to-analog converter. The DAC then outputs an analog signal, in the form of a balancing voltage, to a balance limiting resistor to compensate for bridge zero errors. DAC circuits are sensitive to voltage changes and must be designed to compensate for power supply drifts.

DRIFTING VOLTAGES

In a typical manual balance circuit, Fig. 4-41, it can be shown mathematically that for any static condition of the bridge, the potentiometer can be adjusted for zero volts output. Also, the adjusted zero will be independent of the excitation power supply. Therefore, the supply voltage can change but the gauge output will remain at zero.

For contrast, in a typical DAC circuit the output of the DAC is a function of both the DAC reference voltage and the digital control logic. If the bridge excitation voltage is not the same as that of the DAC reference voltage, the two voltages may drift in opposite directions. This can create large zero errors.

For example, in a system where the bridge excitation is 10 V and the output of the DAC is 5 V, power supply drifts caused by a temperature change of 10 °C can cause an error at

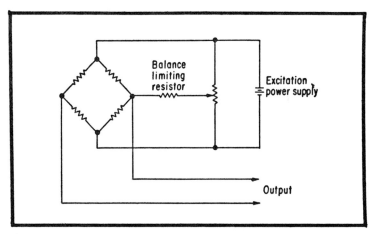

Fig. 4-41. Manual bridge balance circuit uses a potentiometer to calibrate the bridge. This system cannot compensate for amplifier zero drifts.

the output of the bridge of as much as 36 μV. Since strain gauge data signals can be as low as a few microvolts, this can cause significant errors.

Ideally, the reference for the DAC should be derived directly from the voltage across the bridge. This method compensates for power supply drift, and line and load fluctuations. The gauge can be operated from a constant current source, and the DAC will still compensate correctly. The DAC balancing circuit shown in Fig. 4-42 derives its bridge reference in this manner.

The balancing circuit, which calibrates a single strain gauge bridge, can be controlled by a computer. The circuit will accept a zeroing command from the computer, and the strain gauge output can be switched into a multiplexer. A large number of strain gauges, each equipped with this type of circuit can be calibrated in a matter of seconds under computer or operator control.

CIRCUIT DESCRIPTION

Probably the most unique aspect of the circuit is the use of differential amplifiers, Fig. 4-42, to develop the reference voltage for the DAC and to supply the required offset voltage to correct for bridge zero errors.

The reference voltage is derived either from the excitation power supply sense leads, as shown in Fig. 4-42, or from power leads when local sensing or constant current excitation is used. The differential amplifier measures the difference between the two sense lines and ignores any voltage drops due to line loss. It then attenuates and transforms this voltage to the DAC ground reference.

A gain of 0.332 is designed into the reference differential amplifier circuit. This allows the excitation power supply to vary between zero and 20 V, while the output of the amplifier varies from 0.0 to 6.64 V. An upper limit on the output voltage assures that the reference voltage to the DAC will never exceed 7 V. The output of the DAC because of this differential amplifier, is a function of both the digital input word and the bridge excitation voltage.

Output of the DAC feeds the second differential amplifier. This amplifier transforms the voltage from the DAC ground to the excitation supply environment, and has a gain of 3.01. The total system gain, from the reference differential amplifier input to the output of the second differential amplifier, is unity.

The basic balancing circuit shown in Fig. 4-42 contains an Analog Devices' AD7520 10-bit digital-to-analog converter, which provides 1,024 discrete steps. This DAC requires an external reference and interfaces directly with CMOS components. The DAC is controlled by a 12-bit CD4040AE CMOS counter with reset line and an N555T timer.

A three-position switch controls the circuit. The positions are *manual, automatic* and *reset*. In *manual*, the DAC is forced to the center of its output range and the clock is inhibited; this allows a coarse manual bridge balance control to be adjusted. The circuit should be adjusted so that the bridge output is at zero when the excitation supply is set at the normal operating voltage required by the input strain gauge transducer. This provides equal positive and negative correction ranges for the DAC; i.e., it will have an equal number of steps available for correction in either direction. This adjustment needs to be done only once, when the system is initially calibrated.

In the *automatic* position, the circuit is idle, waiting for a reset pulse. The reset pulse can be initiated either remotely, by a computer or remote switch, or by putting the control switch into the momentary reset position. When the reset pulse is received, the counter is cleared to zero, a control latch is reset, and the clock is started.

The clock runs, driving the counter which, in turn, drives the DAC. A comparator circuit compares the output of the

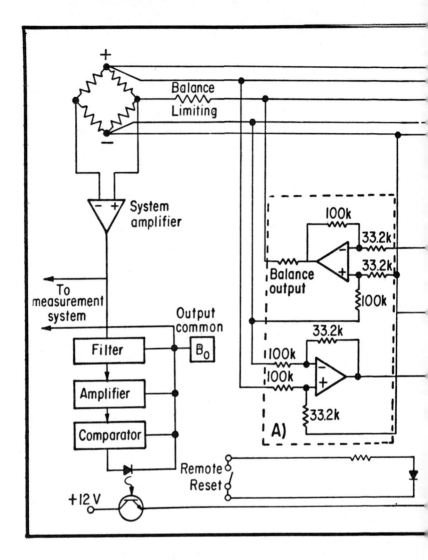

bridge to zero. When the comparator detects zero bridge voltage, it sets the control latch and stops the clock. The counter maintains its last count which keeps the DAC output at a constant voltage. The circuit will continue to maintain the balancing voltage until the next reset pulse is received.

If the circuit cannot find zero, a LED indicator lamp will remain on, and the calibration will have to be accomplished in manual.

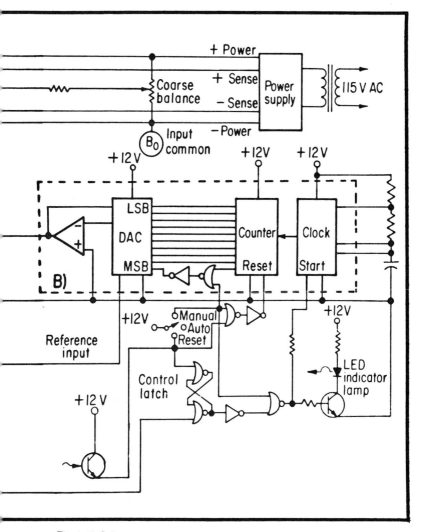

Fig. 4-42. Schematic of a completely automatic bridge balancing system using a DAC. The bridge reference is derived directly from the voltage across the bridge. Differential amplifiers (A) are used to develop the input reference voltage for the DAC and to supply the required offset voltage to the balance limiting resistor. The basic compensation circuit (B) uses a 10-bit DAC, 12-bit CMOS counter and a clock.

The strain gauge differential amplifier output is filtered, amplified and compared to zero by the comparator circuit. The comparator is isolated by an optical coupler which provides the "set" output to the control latch. Other optical couplers maintain complete system isolation between grounds, external switches and the signal conditioning environment.

INCREASING SPEED AND ACCURACY

Higher calibration speeds can be obtained by increasing the clock rate; however, the system time constant determined by the filter must also be increased. Greater accuracy or a wider dynamic range can be accomplished by using a 12-bit DAC. The circuit can also be arranged to correct outputs from other types of transducers. Other features of the circuit include:

- Capable of operating with all types of excitation power supplies, including constant current or constant voltage.
- The circuit maintains system isolation while compensating for bridge errors.
- Remote operation, either computer-controlled or manual, is possible.
- Manual bridge calibrations, if necessary, can easily be accomplished.

JAMES G. STEPHENSON is Director of Engineering for Dynamics Electronic Products, Inc., Chatsworth, CA. Article is based on a paper presented at the ISA Aerospace/Test Measurement Conference, Philadelphia, 1975.

Index

Index

B

C

T